Security and Privacy in Blockchains and the IoT

Security and Privacy in Blockchains and the IoT

Editor

Christoph Stach

MDPI • Basel • Beijing • Wuhan • Barcelona • Belgrade • Manchester • Tokyo • Cluj • Tianjin

Editor
Christoph Stach
University of Stuttgart
Germany

Editorial Office
MDPI
St. Alban-Anlage 66
4052 Basel, Switzerland

This is a reprint of articles from the Special Issue published online in the open access journal *Future Internet* (ISSN 1999-5903) (available at: https://www.mdpi.com/journal/futureinternet/special_issues/SP_BI).

For citation purposes, cite each article independently as indicated on the article page online and as indicated below:

LastName, A.A.; LastName, B.B.; LastName, C.C. Article Title. *Journal Name* **Year**, *Volume Number*, Page Range.

ISBN 978-3-0365-6251-3 (Hbk)
ISBN 978-3-0365-6252-0 (PDF)

Cover image courtesy of Pexels

Contents

About the Editor

Christoph Stach

Dr. rer. nat. Christoph Stach is a postdoctoral researcher at the Applications of Parallel and Distributed Systems department of the University of Stuttgart. He completed his studies in computer science at the University of Stuttgart in 2009. In 2017, he received his PhD in computer science from the University of Stuttgart for his research in the area of information security and data privacy in mobile applications. Following his successful doctorate, he was appointed Academic Councilor at the Institute for Parallel and Distributed Systems of the University of Stuttgart. From June 2020 to September 2021, he held the deputy professorship in Data Engineering at the University of Stuttgart. At present, he is head of the working area of Information Systems and Applications at the Applications of Parallel and Distributed Systems department of the University of Stuttgart. His current research focuses on the concepts and tools required to enable trustworthy and demand-oriented data provisioning for users, such as data scientists and data analysts. To this end, his research addresses research questions regarding data acquisition, data management, data security, and data protection. He has published more than 60 peer-reviewed papers about his research and presented the results at international conferences. For his work, he has received four awards. He also shares his knowledge and experience by giving lectures, such as Introduction to Data Science and Applied Data Science using Python, as well as holding seminars on these topics.

Preface to "Security and Privacy in Blockchains and the IoT"

Smart devices, i.e., everyday objects equipped with comprehensive sensor technology, are becoming increasingly popular. Due to the ubiquity of such devices in our daily lives, data on all kinds of events can continuously be captured and analyzed. As the Internet of Things (IoT) interconnects smart devices, data from a wide range of domains can be linked. Such enriched datasets are the driver for a variety of innovative smart services, e.g., in the eHealth or Industry 4.0 domain. As a result, these data have a high economic value and require special security considerations. Security, in this context, refers to two different facets: On the one hand, the integrity of the data must be protected against illegal manipulation and the availability of the data has to be assured. Blockchain technologies are widely used for this purpose, as they enable the immutable and tamper-resistant storage and sharing of data. On the other hand, applicable data protection laws, such as the EU General Data Protection Regulation (GDPR), set high standards for data processing when it comes to personal data. This is important because a lot of sensitive information can be derived from such data.

To this end, research approaches and insights from practice are discussed in this book, addressing security and privacy issues in the context of blockchain technologies and the IoT. The presented work covers a broad spectrum, ranging from approaches strengthening trust in networks such as the IoT to approaches improving the effectiveness and efficiency of blockchain technologies in terms of query capacities and consensus procedures, as well as approaches enabling the privacy-compliant sharing of IoT data using blockchain technologies. The book is capped off by literature reviews that shed light on what privacy-critical information can be derived from encrypted network traffic flow segments and how blockchain technologies can be leveraged in the ambient assisted-living domain.

As data security and privacy concerns increasingly impact our lives, and blockchain technologies as well as the IoT are prevalent in virtually all domains, the subject matter in this book is, therefore, aimed at both the general and expert audience. The provided overview of the state-of-the-art and state of research addresses developers and researchers as well as end-users. Therefore, this book is recommended to everyone who wants to gain the latest insights and learn about new findings on security and privacy in blockchains and the IoT.

This book has only been made possible due to the authors who have contributed interesting papers about their excellent research work. The editor, therefore, thanks all involved authors. Moreover, he expresses his appreciation to all the reviewers, who were not only essential in selecting the papers for this book but whose valuable comments also ensured that the quality of the selected papers was improved even further. A final word of gratitude goes to the MDPI editorial team, who invested a lot of time and effort in contacting the authors and reviewers and made the publication of this book possible in the first place.

Christoph Stach

Editor

future internet

Editorial

Special Issue on Security and Privacy in Blockchains and the IoT

Christoph Stach

Institute for Parallel and Distributed Systems, University of Stuttgart, Universitätsstraße 38, 70569 Stuttgart, Germany; christoph.stach@ipvs.uni-stuttgart.de

Citation: Stach, C. Special Issue on Security and Privacy in Blockchains and the IoT. *Future Internet* **2022**, *14*, 317. https://doi.org/10.3390/fi14110317

Received: 27 October 2022
Accepted: 28 October 2022
Published: 1 November 2022

Publisher's Note: MDPI stays neutral with regard to jurisdictional claims in published maps and institutional affiliations.

The increasing digitalization in all areas of life is leading step-by-step to a data-driven society. From an information technology perspective, this process is particularly promoted by the Internet of Things (IoT). Nowadays, a variety of sensors can be embedded in virtually any everyday object, enabling users to continuously quantify a wide range of aspects of life. For instance, a smartwatch can use GPS technologies to determine the current location of its user, an accelerometer and gyroscope to recognize the user's activity, and a microphone to capture and interpret voice messages and spoken instructions of its user. Even special use sensors are installed in those IoT devices, such as a heart rate sensor or sensors for recording insulin levels, which can be used to capture and monitor health data. Additionally, such IoT devices have the ability to communicate with each other and exchange the data they gather. In this way, large amounts of data can be collected. Comprehensive processing and analysis of these data (e.g., in a powerful cloud backend) makes it possible to draw conclusions about the context in which an IoT device is used and generate knowledge about the data subjects. This gained knowledge represents the foundation of any smart service, not only in the private sector but also in the public and industrial sectors, such as in the smart home, eHealth, and Industry 4.0 domains.

This renders data as one of the most valuable assets in the information age. Therefore, it is important to manage data securely. Blockchain technologies are often applied to this end, as they ensure the immutability and tamper-resistance of data when they have to be exchanged between multiple parties that do not entirely trust each other. In additions to these information security measures, such highly sensitive data also pose great challenges with respect to data privacy. Applicable data protection laws, such as the EU's General Data Protection Regulation (GDPR), therefore demand the development and application of technical mechanisms that ensure the protection of any natural person when processing their data. In order to ensure that such protective measures are effective, however, they must be tailored to the IoT and blockchain technologies. In this regard, it must be investigated, e.g., how lightweight and privacy-preserving authentication in the IoT is possible; which trust-building approaches regarding the genuineness and validity of IoT data can be applied; and how blockchain systems can efficiently manage big data.

These and related research questions regarding security and privacy in blockchains and the IoT are addressed by six research articles and two literature reviews in this Special Issue. In the following, these eight papers are briefly outlined.

Articles. Two of the research articles address the question of how security and trust in IoT environments and IoT applications can be increased. Alzahrani and Fotiou [1] address how one-to-many communication—or group communication—can be made more secure in software-defined networking (SDN). SDN enables the self-organization of IoT groups by the IoT devices, which reflects the original IoT vision of a network of autonomous things. However, this poses the risk that such an SDN is flooded with fake messages and instructions from malicious things. To counteract this, the authors present an approach in which only authorized endpoints can send instructions to the network. Linked data signatures are used to prove the validity of the instructions. By means of linked data proofs, the presented approach supports zero-knowledge proofs to reliably secure IoT group communications against malicious things. Wei et al. [2] present a different approach to increasing trust in IoT applications. They look at the Social Internet of Things (SIoT), in which smart devices autonomously establish social connections with each other. In this way, whenever necessary, things can become service requesters or service providers on their own, without the need for any human intervention. It is obvious that—similar to real-world services—trust in the service provider is required, e.g., whether it is able to provide the advertised services. While there are approaches that can adequately model this kind of trust in SIoT, these state-of-the-art approaches completely ignore the fact that a service provider has to trust a service requester as well. This work therefore focuses on modeling bidirectional trust in SIoT. This kind of modeling introduces additional complexity due to the fact that trust is context-dependent and can vary depending on the given situation. Based on their bidirectional trust model, the authors discuss a trust-based service delegation method in SIoT, which considers not only the level of trust between a service requester and a service provider but also the utility of the offered service.

Two of the research articles are dedicated to blockchain technologies. While blockchain systems enable secure data management in terms of immutability and tamper resistance, they typically lack comprehensive query capabilities. Przytarski et al. [3] therefore review the current state of query processing in blockchain systems and the future challenges in this research area. For this purpose, they initially investigate in which application domains blockchain technologies are used as part of big data management systems. Based on this, they determine which types of data and which data models are primarily used in this context. They then study the query capabilities of today's blockchain systems and discuss to what extent they meet the requirements of the use cases from the application domains. Furthermore, they give an outlook on how the internal data structures as well as the block structures of a blockchain system have to be adapted in order to efficiently support complex queries, such as history queries over time series data. Qu et al. [4] address another inherent problem in blockchain systems. As the blockchain uses a distributed ledger as its underlying infrastructure, i.e., a replicated, shared, and synchronized data store whose instances are managed by multiple parties, all involved parties have to agree on what data should be added to the blockchain. To synchronize changes, consensus methods such as proof of work (PoW) are used. A major disadvantage of PoW, however, is that it is very computation-intensive and therefore favors parties that have access to powerful computing capacities. In order to provide more fairness in the case of heterogeneous parties, e.g., in IoT environments, the authors interpose edge devices that monitor each computing node participating in PoW. This monitoring is based on a digital twin approach that simulates the normally expected behavior of each computing node. As a result, misbehavior by dishonest participants can be detected, e.g., computing nodes that use extra computing power to outperform their competitors. By means of a proof-of-concept implementation, the authors demonstrate not only the feasibility but also the efficiency of their approach.

The remaining two research articles focus on how blockchain technologies can be applied in the IoT to ensure privacy aspects, namely, access control and privacy-aware data sharing. While IoT applications typically rely on a central data backend that is responsible for the management of the collected data, such an approach poses a risk from a security perspective. Since a single entity operates this backend and thus has full control over the data, tampering is easily possible. Blockchain-based solutions, which manage the data in a distributed manner and are jointly operated by multiple parties, overcome this problem. However, they cause major privacy concerns, as an access policy for confidential data must be reliably applied to all data nodes involved. Khanal et al. [5] therefore introduce a two-pronged approach by which access to sensitive IoT data can only take place with the consent of the data subject. This approach uses a combination of a secure hash function and a key derivation function to encrypt the data. The data in the blockchain can only be decrypted if the data subject has given their consent. With their approach, the authors not only improve reliability but also reduce the computation time compared to state-of-the-art approaches. Gangwani et al. [6] also present an approach with which confidential IoT data can be trustworthily shared among multiple parties using distributed ledger technology. However, their approach relies on IOTA, a distributed-ledger-based communication protocol specifically tailored to the requirements of the IoT. Unlike blockchain-based approaches, IOTA is highly scalable, as restrictions on block size or mining costs are not an issue. As a result, it can also be used to share large amounts of sensor data at a rapid rate. The masked authenticated messaging (MAM) extension for IOTA is used to ensure confidentiality. With MAM, data streams can be sent encrypted as transactions with zero additional cost. Furthermore, MAM provides data subjects with fine-grained access control, allowing them to revoke access to their data at any given time. The authors demonstrate the high potential of IOTA and MAM when dealing with sensitive IoT data by means of an environmental monitoring application.

Reviews. Two literature reviews on in-app activity recognition based on encrypted traffic flow segments and on application areas for blockchain technologies in ambient assisted living wrap up this Special Issue. As the adoption of IoT technologies across all areas of life becomes more and more prevalent, not only the extent of data collection but also the network traffic increases. This is due to the fact that IoT applications do not carry out data processing on the IoT devices themselves, but in a powerful backend. As a result, these applications have to send their data to the backend on a continuous basis. Typically, this data stream is encrypted to ensure that third parties do not gain insight into the transferred payload data. However, even encrypted traffic flow segments still allow conclusions to be drawn about in-app activities, which compromises the privacy of the user. In their review, Pathmaperuma et al. [7] therefore investigate which types of traffic classification exist in the literature. Essentially, there are statistical methods and approaches based on neural networks. In addition to this literature review, the authors also propose their own approach to user activity detection based on in-app data. To this end, they apply an image-based method. Instead of analyzing the network traffic itself, they transform the detected patterns into images, where each pixel stands for features and corresponding feature values of the traffic data. For eight popular mobile applications (e.g., Facebook, Instagram, and WhatsApp), the authors record the network traffic generated by typical in-app activities (e.g., post an image, like an image, and send a short text message). These samples are cleansed, pre-processed, and transformed, resulting in a comprehensive database with characteristic images for each of the in-app activities. A convolutional neural network (CNN) is trained with this image database. The CNN is able to classify activities based on their network traffics with an accuracy of 88 % to 92 %.

One sector that benefits significantly from the IoT and the accompanying comprehensive data collection is the healthcare sector. In particular, recurring routine medical checkups, for instance, in the case of chronic diseases, can be carried out remotely, thus relieving both patients and physicians. As a result of the COVID-19 pandemic, there are increased demands to provide assisted care services remotely. Ambient assisted living (AAL) uses IoT technologies to provide non-intrusive support for the daily lives of elderly or disabled people without the need for a caregiver on site. Since the monitoring required for this purpose collects a large amount of highly personal data, there are justified security and privacy concerns regarding data management. The strategic use of blockchain technologies has the potential to alleviate these concerns. Florea et al. [8] therefore conduct a systematic literature review which aims to identify fields of application for blockchain technologies in the AAL and to highlight advantages and open issues in this context. For this purpose, they selected a literature corpus of 472 scientific papers published in high-quality conferences and journals. In a systematic approach following the PRISM process flow, they condensed this overall corpus to the most relevant papers. Based on these 87 core papers, the authors identify three AAL use cases for which the use of blockchain technologies generates a significant added value. These use cases, which are also further detailed in the review, are IoT-based monitoring and intervention, decentralized patient data management, and AAL system security and privacy. Despite the undeniable benefits that blockchain technologies can provide in these areas, the authors also identify some obstacles that need to be addressed in further research. For instance, there is a need to reduce the transactional and storage costs inherent in managing large amounts of data in blockchain systems, to facilitate the integration of blockchain systems into legacy infrastructures prevalent in many AAL environments, and to ensure the privacy of data managed by a blockchain system.

The eight excellent papers in this Special Issue provide a good overview of security and privacy issues in blockchain systems and the IoT. The research articles present practical solutions to some of these issues. While the literature reviews reveal that there are still several security and privacy issues that need to be addressed in the future, they also show that the use of blockchain technologies and the IoT is beneficial to the daily lives of all of us. It is therefore important to address the questions raised in this Special Issue in the future, in order to make the usage of IoT technologies and blockchain systems as secure and privacy aware as possible.

I would like to thank all the authors for submitting their interesting and informative manuscripts to this Special Issue. I would also like to acknowledge all the reviewers whose thorough and substantial reviews further improved the quality of the manuscripts and without whom this Special Issue would not have been possible. Last but not least, I would like to thank the MDPI editorial team whose support has been instrumental in my work on this Special Issue.

Funding: This research received no external funding.

Conflicts of Interest: The author declares no conflict of interest.

References

1. Alzahrani, B.; Fotiou, N. Securing SDN-Based IoT Group Communication. *Future Internet* **2021**, *13*, 207. [CrossRef]
2. Wei, L.; Yang, Y.; Wu, J.; Long, C.; Lin, Y.B. A Bidirectional Trust Model for Service Delegation in Social Internet of Things. *Future Internet* **2022**, *14*, 135. [CrossRef]
3. Przytarski, D.; Stach, C.; Gritti, C.; Mitschang, B. Query Processing in Blockchain Systems: Current State and Future Challenges. *Future Internet* **2022**, *14*, 1. [CrossRef]
4. Qu, Q.; Xu, R.; Chen, Y.; Blasch, E.; Aved, A. Enable Fair Proof-of-Work (PoW) Consensus for Blockchains in IoT by Miner Twins (MinT). *Future Internet* **2021**, *13*, 291. [CrossRef]
5. Khanal, Y.P.; Alsadoon, A.; Shahzad, K.; Al-Khalil, A.B.; Prasad, P.W.C.; Rehman, S.U.; Islam, R. Utilizing Blockchain for IoT Privacy through Enhanced ECIES with Secure Hash Function. *Future Internet* **2022**, *14*, 77. [CrossRef]
6. Gangwani, P.; Perez-Pons, A.; Bhardwaj, T.; Upadhyay, H.; Joshi, S.; Lagos, L. Securing Environmental IoT Data Using Masked Authentication Messaging Protocol in a DAG-Based Blockchain: IOTA Tangle. *Future Internet* **2021**, *13*, 312. [CrossRef]
7. Pathmaperuma, M.H.; Rahulamathavan, Y.; Dogan, S.; Kondoz, A. CNN for User Activity Detection Using Encrypted In-App Mobile Data. *Future Internet* **2022**, *14*, 67. [CrossRef]
8. Florea, A.I.; Anghel, I.; Cioara, T. A Review of Blockchain Technology Applications in Ambient Assisted Living. *Future Internet* **2022**, *14*, 150. [CrossRef]

future internet

Article

Securing SDN-Based IoT Group Communication

Bander Alzahrani [1] and Nikos Fotiou [2,*]

[1] Faculty of Computing and Information Technology, King Abdulaziz University, Jeddah 21589, Saudi Arabia; baalzahrani@kau.edu.sa
[2] Mobile Multimedia Laboratory, Department of Informatics, School of Information Sciences and Technology, Athens University of Economics and Business, Patision 76, 10434 Athens, Greece
* Correspondence: fotiou@aueb.gr

Abstract: IoT group communication allows users to control multiple IoT devices simultaneously. A convenient method for implementing this communication paradigm is by leveraging software-defined networking (SDN) and allowing IoT endpoints to "advertise" the resources that can be accessed through group communication. In this paper, we propose a solution for securing this process by preventing IoT endpoints from advertising "fake" resources. We consider group communication using the constrained application protocol (CoAP), and we leverage Web of Things (WoT) Thing Description (TD) to enable resources' advertisement. In order to achieve our goal, we are using linked-data proofs. Additionally, we evaluate the application of zero-knowledge proofs (ZKPs) for hiding certain properties of a WoT-TD file.

Keywords: crowd management; software-defined networking; linked-data signatures; Web of Things; zero-knowledge proofs

Citation: Alzahrani, B.; Fotiou, N. Securing SDN-Based IoT Group Communication. *Future Internet* **2021**, *13*, 207. https://doi.org/10.3390/fi13080207

Academic Editor: Christoph Stach

Received: 9 July 2021
Accepted: 3 August 2021
Published: 9 August 2021

Publisher's Note: MDPI stays neutral with regard to jurisdictional claims in published maps and institutional affiliations.

1. Introduction

The Internet of Things (IoT) considers "unconventional" communication paradigms such as "publish–subscribe" or "one-to-many" communication; in this paper, we focus on the latter paradigm, which is usually referred to as group communication. Although this paradigm is not typical in mainstream communication systems, we postulate that this is not the case for the IoT. We consider as a use case the crowd monitoring system presented in [1]. This system includes gas and ultrasonic sensors, UAVs equipped with cameras and LiDARs, as well as CCTV systems (see also Figure 1). In this system, tasks such as "collect all measurements in area X" or "turn on all cameras in area Z" are not uncommon scenarios, and the reasonable approach for implementing them is using group communication.

Group communication using the constrained application protocol (CoAP) [2] is a promising direction, which is impeded by the lack of adoption of IP Multicast, however. On the other hand, interconnecting IoT devices over software-defined networking (SDN)—such as in the architecture presented in [1]—enables alternative approaches for implementing group communication that removes the need for IP multicast and enables "self-organizing" IoT groups, where IoT endpoints can "advertise" the CoAP URIs of their resources, and groups can be automatically created based on these advertisements. Therefore, it is obvious that this advertisement process must be protected; otherwise, malicious entities may "pollute" the network with "fake" advertisements, affecting this way the group formation process.

In this paper, we provide a solution to this problem by allowing only authorized endpoints to perform advertisements. From a high-level perspective, we consider that each endpoint "represents" its available resource using a JSON-encoded file and by following the W3C Web of Things, Thing Description (WoT-TD) specifications [3]. This WoT-TD file is signed by a trusted service provider, and it is included in the advertisements, together with proof of ownership. The recipients of such an advertisement can then easily verify its validity. In this paper, we make the following contributions:

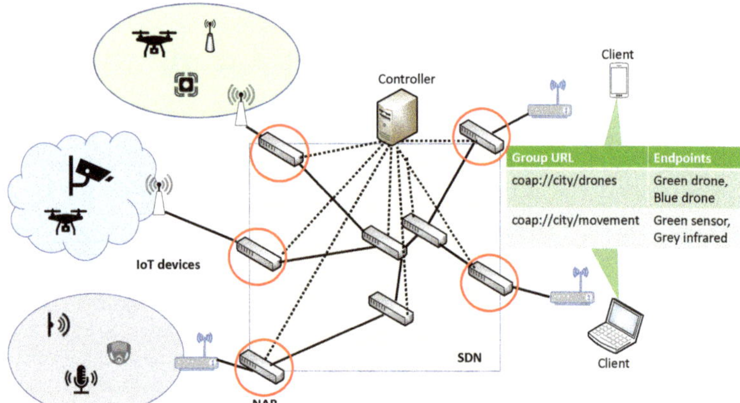

Figure 1. An overview of the entities of the proposed solution.

- We design an IoT onboarding process that ensures that only authorized IoT devices participate in a group;
- We leverage linked-data signatures to provide advertisement validity and proof of ownership;
- We extend our solution to support selective advertisement of resources using zero-knowledge proofs.

The remainder of this paper is organized as follows: In Section 2, we introduce the technologies used as building blocks of our system. In Section 3, we detail the design of our solution, and in Section 4, we present its implementation and evaluation. We discuss related work in this area in Section 5, and we conclude our paper in Section 6.

2. Background

2.1. SDN and Bloom Filter-Based Forwarding

Software-defined networking (SDN) [4] is a technology that allows a centralized entity, known as the "network controller" (or simply controller) to control programmable switches. SDN switches forward packets based on rules defined by the controller. In particular, in order for a switch to determine how to handle an incoming packet, it either queries the controller or uses a set of rules stored in the switch using a protocol such as OpenFlow [5].

SDN can be used for implementing Bloom filter [6]-based packet forwarding [7]. This type of forwarding enables multicast communication in an efficient, fast, and stateless way. From a high-level perspective, the solution in [7] assumes that each outgoing interface of an SDN switch is identified by a bitstring identifier; then, it uses a Bloom filter to encode the identifiers of all interfaces through which a packet should be forwarded; finally, it stores this forwarding identifier in the IPv6 address field of the packet. SDN switches are preconfigured with rules that allow them to decide the outgoing interface of each incoming packet simply by "ORing" the packet's forwarding identifier with the identifiers of all outgoing interfaces.

2.2. CoAP and CoAP Group Communication

CoAP [8] is a lightweight protocol, designed to be the "HTTP of the IoT." CoAP resources are identified by a URI scheme, similar to HTTP URIs, and the CoAP interaction model is similar to the client–server model of HTTP. Therefore, IoT endpoints act as CoAP "servers", exposing one or more CoAP URIs that can be accessed by CoAP "clients" using a suitable CoAP "method" .

CoAP group communication is a CoAP extension that allows CoAP clients to retrieve (or set) resources from a group of CoAP servers, e.g., retrieve the temperature measurements from all sensors of a building, turn on and off all the lights of a smart city, etc. An approach for realizing CoAP group communication is using IP multicast (Section 2 of [2]). With this approach, CoAP servers belonging to the same group join an IP multicast address, and CoAP clients learn the IP multicast address of a group using DNS resolution. Then, CoAP clients can send CoAP requests to an IP multicast address and receive the corresponding response(s) using unicast. Nevertheless IP multicast is not the only option; other underlay networking architectures can be used instead. For example, as as we discuss in the following section, our solution relies on SDN to implement one-to-many communication.

2.3. Web of Things

The goal of W3C's Web of Things (WoT) working group is to improve the interoperability and usability of the Internet of Things (IoT) [9] by specifying universal means for accessing IoT devices. This goal is achieved by providing building blocks that leverage and extend existing, standardized Web technologies in the context of IoT. Such a building block is the Thing Description specification draft [3].

A Thing Description (WoT-TD) is a JSON-LD [10] document that describes the "metadata" and "interfaces" of a "thing", where a thing can be a physical IoT device or a virtual entity that is composed of multiple IoT devices. An interface can be a "property", an "action", or an "event", that can be accessed using a Web technology such as CoAP or HTTP. A WoT-TD describes how these interfaces can be accessed by specifying suitable URIs, security policies, and other information that can be used by an interested client.

Being encoded using JSON-LD, a WoT-TD includes a `context` property, which is an array of URLs pointing to documents that include "vocabulary" terms. All WoT-TD include the "https://www.w3.org/2019/wot/td/v1" context, but additional contexts can be added, allowing the extension of the WoT-TD vocabulary (see also Section 7.1 of [3] for more information).

3. Overview

3.1. System Entities and Security Assumptions

Our solution considers an SDN network that interconnects IoT *endpoints* acting as CoAP servers with IoT service clients. Each IoT endpoint owns an Ed22519 public key [11], denoted by $Endpoint_{ID}$. Network operators maintain a list of $Endpoint_{ID}$ identifiers belonging to authorized IoT devices. IoT devices are connected to the SDN network through an access device; the type of this device depends on the available communication technology (e.g., WiFi, ZigBee, LoRa, etc.); nevertheless, our solution is oblivious to used technology. Access devices are connected to edge switches acting as the network attachment point (NAP), and it is assumed that there are mechanisms that allow NAPs to access this list of authorized IoT devices. Therefore, it should be not possible for an attacker to join a network by impersonating an authorized IoT device inside the SDN network. Similarly, the network operator owns a well-known Ed22519 public key, denoted by $Network_{ID}$, which acts as the root of trust in our system; all endpoints are preconfigured with this key, and all NAPs can generate signatures using the private key that corresponds to a $Network_{ID}$. Whenever a key is included in a text-based file, we are using its Base64url encoding [12].

IoT devices provide access to resources. We focus on resources that can be accessed using CoAP and CoAP group communication.

Additionally, we consider that the SDN controller knows the full network topology, and it is capable of constructing forwarding paths from an $Endpoint_{ID}$ toward one or more $Endpoint_{ID}$ identifiers (see also Section 2.1). These paths are identified by a Bloom filter-based identifier denoted as Fwd_{ID}. Finally, we consider that there is a well-known Fwd_{ID} that can be used by endpoints to broadcast packets in the network.

Our solution is focused on IoT endpoints acting as CoAP servers; for this reason, we neither consider clients as part of our threat model nor are we concerned with client-facing security operations such as client authentication and authorization.

3.2. IoT Device Onboarding

Each IoT device is preconfigured with a WoT-TD file for the resources it provides. An example of a WoT-TD is one that includes an ultrasonic sensor deployed in a stadium, which can be seen in Listing 1. Lines 1–5 define the *context* of the WoT-TD file and include the identifier of the IoT device, i.e., the $Endpoint_{ID}$. This example also includes an *action* used for "turning off" the sensor (line 8). This action can be invoked using either plain CoAP (lines 10–13) or CoAP group communication (lines 14–23). As it can be observed, this action can be invoked through two different groups, one representing "all sensors of the stadium" (line 15) and another representing "all sensors of a city" (line 20).

Listing 1. An example of a WoT-TD file.

```
1     {
2       "context": "https://www.w3.org/.../v1",
3       ...
4       "id": EndpointID,
5       }
6       "properties": {...},
7       "actions": {
8         "turnoff_sensor": {
9           "forms":
10            [{
11               "href":"coap://gate7.stadium/sensor1/turnoff",
12               "cov:methodName": "POST"
13            },
14            {
15               "href":"coap://stadium/sensors/turnoff",
16               "cov:methodName": "POST"
17               "subprotocol":"cov:group"
18            },
19            {
20               "href":"coap://city1/sensors/turnoff",
21               "cov:methodName": "POST"
22               "subprotocol":"cov:group"
23            }]
24         }
25       },
26       "events": {...}
27     }
```

In order for an IoT device to join the network, it establishes a (D)TLS communication channel with a NAP. The (D)TLS handshake uses the "client authentication" option. The goal of this handshake is to allow the IoT device to verify that the NAP knows the private key that corresponds to $Network_{ID}$, and the NAP to verify that the IoT device is the owner of $Endpoint_{ID}$.

As a next step, the IoT device sends its WoT-TD file to the NAP, and the NAP verifies that it includes the same $Endpoint_{ID}$ used during the handshake, as well as that the $Endpoint_{ID}$ is in the list of authorized identifiers. Then, the NAP signs the WoT-TD file using a *linked-data proof* (LDP) [13]. An LDP is a mechanism for ensuring the authenticity and integrity of linked-data documents, such as WoT-TD files, which is extensible and supports contemporary cryptographic solutions, such as zero-knowledge proofs (ZKPs). An LDP is encoded using JSON, and an example of an LDP used in our system is included in Listing 2. As it can be seen, line 2 defines the type of the proof, which, in our example, is an EdDSA signature [11]; line 3 includes a timestamp indicating the proof creation time; line 4 includes information that can be used for verifying the proof, which, in our case, is the $Network_{ID}$; line 5 includes the purpose of the proof, which, in this listing, is to provide an "assertion" about the integrity of the WoT-TD file; finally, line 6 is the actual proof value.

Listing 2. A linked-data proof used in our system.

```
1  {
2      "type": "Ed25519Signature2020",
3      "created": "2021-17-06T11:01:24Z",
4      "verificationMethod": Network_ID,
5      "proofPurpose": "assertionMethod",
6      "proofValue": "VqpLMweBrSxMY2x...aqA3Q1geV6"
7  }
```

The received proof is appended to the WoT-TD file. From this point on, the IoT device can participate in the rest of the operations of our system.

3.3. SDN-Based IoT Group Communication

Our system adapts the solution presented in [14] for providing SDN-based IoT group communication and implements IoT group communication as a two-step process. The first step involves the *advertisement* of the available resources, and the second step implements the actual CoAP *group requests*.

3.3.1. Resource Advertisement

IoT devices should advertise their resources by broadcasting their WoT-TD files. As a reminder, we assume a well-known Fwd_{ID} that can be used for broadcasting. In order to protect advertisements from replay attacks, IoT devices generate a new LDP, similar to the one they have received by the NAP, which, however, includes a *nonce*, and is generated using the private key that corresponds to $Endpoint_{ID}$. Nonces in our system must not be reused within a specific time frame. This can be easily implemented by maintaining a list of used nonces: each nonce should remain in the list for the duration of the selected time frame, and devices must make sure that a nonce they send or receive is not included in that list.

Upon receiving an advertisement, clients validate the integrated LDPs. In particular, they validate that the advertisement includes an LDP that can be verified using $Network_{ID}$, which is "well known", and another that can be verified using $Endpoint_{ID}$ included in the WoT-TD file. Additionally, they verify that the latter proof is adequately fresh, and it includes a unique nonce. If all verifications are successful, each client updates a *lookup table* that includes mappings from CoAP group URIs to the corresponding $Endpoint_{ID}$ identifiers.

3.3.2. CoAP Group Request

A CoAP client wishing to send a request to a CoAP group implements the related protocol described in [14]. From a high-level perspective the client executes the following steps:

1. From the lookup table, it retrieves the $Endpoint_{ID}$ identifiers of the CoAP servers that are members of the group;
2. If it knows a Fwd_{ID} for all retrieved $Endpoint_{ID}$ identifiers, it proceeds to step 4;
3. It constructs a message that includes all $Endpoint_{ID}$ identifiers for which it does not know a Fwd_{ID} and sends it to a "special" MAC address used for forcing SDN switches to forward a packet to the controller. The controller responds with a list of Fwd_{ID} that is eventually returned back to the client;
4. It creates a new Fwd_{ID} by ORing the Fwd_{ID} identifiers of the retrieved $Endpoint_{ID}$ identifiers and forwards the CoAP request using the created Fwd_{ID}. Due to the properties of Bloom filter-based forwarding (see [7] for more details), the CoAP request will be forwarded to the appropriate IoT devices.

4. Implementation and Evaluation

4.1. Implementation

We implemented our solution using Eclipse's Thingweb node-wot (https://github.com/eclipse/thingweb.node-wot accessed on 5 August 2021) as an IoT endpoint, and libcoap library (https://libcoap.net/ accessed on 5 August 2021) for emulating CoAP clients. For the SDN underlay, we relied on the tools presented in [15], i.e., we used the Open vSwitch [16] SDN switch, the POX [17] SDN controller, and we emulated the network using the mininet network emulator [18]. Finally, we used the JSON-LD library (https://github.com/digitalbazaar/jsonld-signatures accessed on 5 August 2021) to generate and verify LDPs.

In order to not modify libcoap to support the used SDN-based group communication, we developed a CoAP proxy that implements the related protocols; CoAP clients wishing to send a request to a group simply forward their request to the proxy using plain CoAP (see Section 5.7 of [8]). Using this approach, group communication is implemented transparently from the used CoAP library. This is a useful property since it allows the use of our solution even with constrained IoT devices, acting as a CoAP client, although using CoAP libraries with limited functionality.

We measured the time required to generate and verify an LDP in a desktop PC equipped with an Intel-i5 CPU and 4GB RAM, running Xubuntu, and a Raspberry Pi 2 Model B Rev 1.1 with a 900 MHz quad-core ARM Cortex-A7 CPU and 1GB RAM, running Raspberry Pi OS. Table 1 shows the results. The size of the corresponding base64-encoded LDP is 508 bytes.

Table 1. LDP generation and verification times.

Operation	Desktop	Raspberry Pi
LDP generation	0.93 ms	5.2 ms
LDP verification	0.83 ms	6.0 ms

4.2. Security Evaluation

The security goal of our solution is to prevent "fake" advertisements. Indeed, with our solution, only authorized IoT endpoints can advertise WoT-TD files. Furthermore, because these WoT-TD files are singed, neither an active attacker nor the IoT endpoint itself can modify them.

An active attacker in our system is able to replay valid advertisements. Although, in general, replay attacks are prevented by the use of the *nonce*, there can be cases in which the replayed advertisement is received before the real one. In these cases, an endpoint will believe that the attacker is a legitimate IoT device and that the real advertisement was a replayed one. Although this attack cannot be prevented in a straightforward way, we argue that its impact is limited, and it can be easily detected and i mitigated. Advertisements in our system do not contain any location-specific information, since $Endpoint_{ID}$ identifiers are just public keys. Therefore, if an $Endpoint_{ID}$ still provides the advertised resources, the attack will have no impact apart from the added network overhead. Moreover, advertisements in our system are broadcasted; hence, it will be trivial for a monitoring entity to detect the replay attack. Finally, we consider an SDN-based architecture, in which a controller can remove any endpoint from the network in a straightforward manner.

Similarly, an attacker that has access to the private key that corresponds to an $Endpoint_{ID}$ can only sent *valid* advertisements of WoT-TD files belonging to the corresponding IoT device, i.e., since the $Endpoint_{ID}$ is included in the WoT-TD, the attacker cannot use the breached key to sign an advertisement of another IoT device. Therefore, the impact of this attack is similar to the impact of the replay attack.

From a security perspective, the most critical component of our solution is the private key that corresponds to $Network_{ID}$. If this key is compromised, then it must be revoked; hence, all endpoints must be reconfigured with the new $Network_{ID}$, and all LDPs must be regenerated.

4.3. Private Advertisements Using ZKPs

Zero-knowledge proofs (ZKP) are a class of proofs in which a *prover* can prove to a *verifier* the knowledge of a value without revealing what the value is [19,20]. In the context of our system, ZKPs can be used by a user in order to generate a WoT-TD that reveals only affordances that are accessed through CoAP group communication. In order to achieve this, the proof of the TD signature must have been generated using an appropriate signature algorithm. In our implementation, we used the BBS+ linked signature algorithm for this purpose [21]. This signature scheme makes use of BLS12-381 pairing-friendly keys [22]. In a WoT-TD that contains a BBS+ signature, a subset of the affordances can be hidden by "framing" the original WoT-TD in a JSON-LD frame [23]. A *JSON-LD frame* can be seen as a filter that, when applied to a JSON-LD document (e.g., a W3C-compliant VC), it outputs a new JSON-LD document that contains only a subset of the fields of the original document. Table 2 shows the time required to generate and verify an LDP proof, using the same endpoints as in Section 4.1, for a WoT-TD file that includes four affordances, three of which are hidden. For this purpose, we used node.js and the jsonld-bbs library, (https://github.com/mattrglobal/jsonld-signatures-bbs accessed on 5 August 2021), which handles JSON-LD objects and uses BLS12-381 keys to generate BBS+ ZKPs. The size of the corresponding base64-encoded LDP is 891 bytes.

Table 2. LDP generation and verification times when BBS+ is used.

Operation	Desktop	Raspberry Pi
LDP	174.6 ms	999.2 ms
LDP verification	74.3 ms	1004.0 ms

As can be seen from this Table, LDP generation and verification in the Raspberry Pi requires approximately 1 s. Nevertheless, we are using an old device and an implementation written in node.js; therefore, there is significant space for improvement. Additionally, these signatures have to be calculated every time a new WoT description file is advertised; this process does not have to take place often—its frequency can be in the order of hours.

5. Related Research

Many research efforts provide solutions for protecting the confidentiality of IoT group communication messages, e.g., using group object security for constrained restful environments (OSCORE) [24], DTLS with pre-shared keys among group members [25,26], attribute-based encryption [27], or even by relying on the information-centric networking (ICN) paradigm [28]. These solutions are concerned with the establishment of a secret key that is used for encrypting data [24,28] or the communication channels [25,26]. Such a key can be derived by a pre-shared symmetric key, a public key, or by the attributes of the communicating endpoints. Our solution is orthogonal to these approaches since our goal is to make sure that a client receives CoAP responses only from authorized servers.

Our work considers an SDN-based underlay architecture used instead of IP multicast for implementing group communication. Many related efforts are considering other alternatives to IP multicast, including the use of bit index explicit replication (BIRE) [29], MPL [30], and ICN [31]. We see these efforts complementary to our approach since our solution is agnostic to the underlying mechanism; therefore, it could be used with any of them.

Some related solutions use identity-based signature (IBS) to achieve similar goals with our work (see, for example, [32]). Although IBS removes the need for public keys, it introduces computational overhead, and it suffers from the so-called key escrow problem, since there is a single entity (the key generator) that knows all private keys. Our solution uses Ed25519 keys, which are 32 bytes long; hence, the gains, in terms of communication overhead of using an identity rather than an Ed25519 key, are small.

Our solution is designed for IoT devices and gateways that support the WoT specification. However, in recent years, a number of related technologies have emerged. For example, under the umbrella of the *European IoT Platform Initiative* (https://iot-epi.eu/ accessed on 5 August 2021) a number of IoT gateway technologies were developed, by projects such as symbIoTe (https://iot-epi.eu/project/symbiote/ accessed on 5 August 2021), AGILE (https://iot-epi.eu/project/agile/ accessed on 5 August 2021), and Interiot (https://iot-epi.eu/project/inter-iot/ accessed on 5 August 2021). These efforts are now stalled. Since the only requirement of our solution is that device descriptions are encoded using JSON, we believe that it can be easily adapted for other gateway technologies.

Our system uses public keys for identifying IoT endpoints. A relevant technology that can be used instead is that of *decentralized identifiers* (DIDs) [33]. DIDs are closely related to LDPs. Additionally, when applied to the IoT, DIDs have some interesting security and privacy properties [34].

6. Conclusions

In this paper, we presented a security solution for managing group membership in IoT group communication. In particular, we leveraged linked-data proofs to assure that only valid group members can "advertise" their available resources. Our solution has intriguing security properties since it is a resilient event to private key breaches. Linked-data proofs allow the use of zero-knowledge proofs; our solution leverages this property in order to implement "selective disclosure" of available resources.

Future work in this area includes the integration of our solution with content confidentiality mechanisms (e.g., group OSCORE).

Author Contributions: Investigation, B.A. and N.F.; Project administration, B.A.; Software, N.F.; Writing—review & editing, B.A. and N.F. All authors have read and agreed to the published version of the manuscript.

Funding: This research was funded by Deputyship for Research & Innovation, Ministry of Education in Saudi Arabia grant number 277.

Data Availability Statement: Not Applicable, the study does not report any data.

Acknowledgments: The authors extend their appreciation to the Deputyship for Research & Innovation, Ministry of Education in Saudi Arabia for funding this research work through the project number (227).

Conflicts of Interest: The authors declare no conflict of interest.

References

1. Jiang, Y.; Miao, Y.; Alzahrani, B.; Barnawi, A.; Alotaibi, R.; Hu, L. Ultra Large-Scale Crowd Monitoring System Architecture and Design Issues. *IEEE Internet Things J.* **2021**, *8*, 10356–10366. [CrossRef]
2. Rahman, A.; Dijk, E. Group Communication for the Constrained Application Protocol (CoAP). 2014. Available online: https://www.hjp.at/doc/rfc/rfc7390.html (accessed on 5 August 2021).
3. Kaebish, S.; Kamiya, T.; McCool, M.; Charpenay, V.; Kovatsch, M. Web of Things Thing Description. 2020. Available online: https://www.w3.org/TR/wot-binding-templates/ (accessed on 5 August 2021).
4. Xia, W.; Wen, Y.; Foh, C.H.; Niyato, D.; Xie, H. A Survey on Software-Defined Networking. *IEEE Commun. Surv. Tutor.* **2015**, *17*, 27–51. [CrossRef]
5. Lara, A.; Kolasani, A.; Ramamurthy, B. Network Innovation using OpenFlow: A Survey. *IEEE Commun. Surv. Tutor.* **2014**, *16*, 493–512. [CrossRef]
6. Bloom, B.H. Space/time trade-offs in hash coding with allowable errors. *Commun. ACM* **1970**, *13*, 422–426. [CrossRef]
7. Reed, M.J.; Al-Naday, M.; Thomos, N.; Trossen, D.; Petropoulos, G.; Spirou, S. Stateless multicast switching in software defined networks. In Proceedings of the 2016 IEEE International Conference on Communications (ICC), Kuala Lumpur, Malaysia, 22–27 May 2016; pp. 1–7. [CrossRef]
8. Shelby, Z.; Hartke, K.; Bormann, C. The Constrained Application Protocol (CoAP). 2014. Available online: https://iottestware.readthedocs.io/en/master/coap_rfc.html (accessed on 5 August 2021).
9. Kovatsch, M.; Matsukura, R.; Lagally, M.; Kawaguchi, T.; Toumura, K.; Kajimoto, K. Web of Things Architecture. 2020. Available online: https://www.w3.org/TR/wot-architecture/ (accessed on 5 August 2021).
10. Kellogg, G.; Champin, P.; Longley, D. JSON-LD 1.1. 2020. Available online: https://www.w3.org/TR/json-ld11/ (accessed on 5 August 2021).

11. Bernstein, D.J.; Duif, N.; Lange, T.; Schwabe, P.; Yang, B.Y. High-speed high-security signatures. *J. Cryptogr. Eng.* **2012**, *2*, 77–89. [CrossRef]

12. Josefsson, S. The Base16, Base32, and Base64 Data Encodings. 2006. Available online: https://www.hjp.at/doc/rfc/rfc4648.html (accessed on 5 August 2021).

13. Longley, D.; Sporny, M. Linked Data Proofs 1.0. 2021. Available online: https://w3c-ccg.github.io/ld-proofs/ (accessed on 5 August 2021).

14. Fotiou, N.; Mendrinos, D.; Polyzos, G.C. Edge-assisted traffic engineering and applications in the IoT. In Proceedings of the 2018 Workshop on Mobile Edge Communications (MECOMM'18), Budapest, Hungary, 20–25 August 2018; Association for Computing Machinery: New York, NY, USA, 2018; pp. 37–42.

15. Alzahrani, B.; Fotiou, N. Enhancing Internet of Things Security using Software-Defined Networking. *J. Syst. Archit.* **2020**, *110*, 101779. [CrossRef]

16. Pfaff, B.; Pettit, J.; Koponen, T.; Jackson, E.; Zhou, A.; Rajahalme, J.; Gross, J.; Wang, A.; Stringer, J.; Shelar, P.; et al. The design and implementation of open vSwitch. In Proceedings of the 12th USENIX Symposium on Networked Systems Design and Implementation (NSDI 15), Oakland, CA, USA, 4–6 May 2015; USENIX Association: Oakland, CA, USA, 2015; pp. 117–130.

17. Gude, N.; Koponen, T.; Pettit, J.; Pfaff, B.; Casado, M.; McKeown, N.; Shenker, S. NOX: Towards an Operating System for Networks. *SIGCOMM Comput. Commun. Rev.* **2008**, *38*, 105–110. [CrossRef]

18. Lantz, B.; Heller, B.; McKeown, N. A network in a laptop: Rapid prototyping for software-defined networks. In Proceedings of the 9th ACM SIGCOMM Workshop on Hot Topics in Networks, Monterey, CA, USA, 20–21 October 2010; ACM: New York, NY, USA, 2010; pp. 19:1–19:6.

19. Goldreich, O.; Oren, Y. Definitions and properties of zero-knowledge proof systems. *J. Cryptol.* **1994**, *7*, 1–32. [CrossRef]

20. Quisquater, J.J.; Quisquater, M.; Quisquater, M.; Quisquater, M.; Guillou, L.; Guillou, M.A.; Guillou, G.; Guillou, A.; Guillou, G.; Guillou, S. How to explain zero-knowledge protocols to your children. In Proceedings of the Conference on the Theory and Application of Cryptology, Santa Barbara, CA, USA, 20–24 August 1989; pp. 628–631.

21. Looker, T.; Steele, O. BBS+ Signatures 2020. 2020. Available online: https://w3c-ccg.github.io/ldp-bbs2020/ (accessed on 5 August 2021).

22. Sakemi, Y. (Ed.) Pairing-Friendly Curves v09. 2020. Available online: https://datatracker.ietf.org/doc/html/draft-irtf-cfrg-pairing-friendly-curves-09 (accessed on 5 August 2021).

23. W3C. JSON-LD 1.1 Framing. 2020. Available online: https://www.w3.org/TR/json-ld11/ (accessed on 16 July 2020).

24. Tiloca, M.; Selander, G.; Palombini, F.; Park, J. Group OSCORE—Secure Group Communication for CoAP. 2021. Available online: https://www.ietf.org/id/draft-ietf-core-oscore-groupcomm-12.html (accessed on 5 August 2021).

25. Tiloca, M.; Nikitin, K.; Raza, S. Axiom: DTLS-Based Secure IoT Group Communication. *ACM Trans. Embed. Comput. Syst.* **2017**, *16*, 1–29. [CrossRef]

26. Park, C.S. Security Architecture for Secure Multicast CoAP Applications. *IEEE Internet Things J.* **2020**, *7*, 3441–3452. [CrossRef]

27. Basu, S.S.; Tripathy, S. Securing Multicast Group Communication in IoT-Enabled Systems. *IETE Tech. Rev.* **2019**, *36*, 83–93. [CrossRef]

28. Gündoğan, C.; Amsüss, C.; Schmidt, T.C.; Wählisch, M. IoT content object security with OSCORE and NDN: A first experimental comparison. In Proceedings of the 2020 IFIP Networking Conference (Networking), Paris, France, 22–26 June 2020; pp. 19–27.

29. Wijnands, I.; Rosen, E.C.; Aldrin, S.; Przygienda, T.; Dolganow, A. Multicast Using Bit Index Explicit Replication (BIER). 2017. Available online: https://www.hjp.at/doc/rfc/rfc8279.html (accessed on 5 August 2021).

30. Hui, J.; Kelsey, R. M Multicast Protocol for Low-Power and Lossy Networks (MPL). 2016. Available online: https://tex2e.github.io/rfc-translator/html/rfc7731.html (accessed on 5 August 2021).

31. Gündoğan, C.; Amsüss, C.; Schmidt, T.C.; Wählisch, M. Toward a restful information-centric web of things: A deeper look at data orientation in CoAP. In Proceedings of the 7th ACM Conference on Information-Centric Networking (ICN'20), Virtual, 29 September–1 October 2020; Association for Computing Machinery: New York, NY, USA, 2020; pp. 77–88.

32. Felde, N.; Grundner-Culemann, S.; Guggemos, T. Authentication in dynamic groups using identity-based signatures. In Proceedings of the 2018 14th International Conference on Wireless and Mobile Computing, Networking and Communications (WiMob), Limassol, Cyprus, 15–17 October 2018; pp. 1–6. [CrossRef]

33. W3C Credentials Community Group. A Primer for Decentralized Identifiers. 2019. Available online: https://w3c-ccg.github.io/did-primer/ (accessed on 5 August 2021).

34. Kortesniemi, Y.; Lagutin, D.; Elo, T.; Fotiou, N. Improving the Privacy of IoT with Decentralised Identifiers (DIDs). *J. Comp. Netw. Commun.* **2019**, *2019*, 8706760:1–8706760:10. [CrossRef]

 *future internet*

Article

A Bidirectional Trust Model for Service Delegation in Social Internet of Things

Lijun Wei [1], Yuhan Yang [1], Jing Wu [1], Chengnian Long [1,*] and Yi-Bing Lin [2,3,*]

[1] Department of Automation, Shanghai Jiao Tong University, Shanghai 200240, China; sjtu_weilijun@sjtu.edu.cn (L.W.); yuhanyang@sjtu.edu.cn (Y.Y.); jingwu@sjtu.edu.cn (J.W.)
[2] Department of Computer Science, National Yang Ming Chiao Tung University, Hsinchu 30010, Taiwan
[3] College of Humanities and Sciences, China Medical University, Taichung 406, Taiwan
[*] Correspondence: longcn@sjtu.edu.cn (C.L.); liny@csie.nctu.edu.tw (Y.-B.L.)

Abstract: As an emerging paradigm of service infrastructure, social internet of things (SIoT) applies the social networking aspects to the internet of things (IoT). Each object in SIoT can establish the social relationship without human intervention, which will enhance the efficiency of interaction among objects, thus boosting the service efficiency. The issue of trust is regarded as an important issue in the development of SIoT. It will influence the object to make decisions about the service delegation. In the current literature, the solutions for the trust issue are always unidirectional, that is, only consider the needs of the service requester to evaluate the trust of service providers. Moreover, the relationship between the service delegation and trust model is still ambiguous. In this paper, we present a bidirectional trust model and construct an explicit approach to address the issue of service delegation based on the trust model. We comprehensively consider the context of the SIoT services or tasks for enhancing the feasibility of our model. The subjective logic is used for trust quantification and we design two optimized operators for opinion convergence. Finally, the proposed trust model and trust-based service delegation method are validated through a series of numerical tests.

Keywords: trust model; social internet of things; service delegation

Citation: Wei, L.; Yang, Y.; Wu, J.; Long, C.; Lin, Y.-B. A Bidirectional Trust Model for Service Delegation in Social Internet of Things. *Future Internet* **2022**, *14*, 135. https://doi.org/10.3390/fi14050135

Academic Editor: Christoph Stach

Received: 9 April 2022
Accepted: 26 April 2022
Published: 29 April 2022

Publisher's Note: MDPI stays neutral with regard to jurisdictional claims in published maps and institutional affiliations.

1. Introduction

As the 4th industrial revolution and the development of future social interconnection technology, internet of things (IoT), following the internet, brings tremendous changes in people's lives [1–3]. With the continuous intelligence of hardware devices and the maturity of edge computing technology, IoT will have greater scalability [4,5]. Integrating the concept of socialization into the IoT system, the social internet of things (SIoT) [6,7], as a new service paradigm, improves the interoperability among IoT objects and enhances the service efficiency in industry applications. The objects will establish the relationship with each other and collaborate on services without human intervention, which make the objects more autonomous in the process of IoT service. Moreover, the structure of SIoT boosts the network navigability and scalability, which enhances the service discovery and resource acquisition. Currently, the SIoT paradigm has been widely applied in various application scenarios, such as vehicular social networks [8–11], mobile crowdsensing [12–16], data-driven smart city [17–20], etc.

In SIoT, each object (e.g., intelligent sensors, smartphone, and video camera) can be a service requester (SR) or service provider (SP), according to its own motivations. The SR will broadcast the service request, such as collecting sensing tasks or urban noise data, and provide some rewards to the SP. On the other hand, the SP will provide the specific service, such as sharing information or computation resources to the SR, to receive some rewards from the SR. Each IoT object can autonomously determine which service to initiate and which object to delegate within a given set of candidate objects. By this method, the service discovery, interaction, and execution will be optimally implemented.

Although the SIoT paradigm will improve the quality of services to a certain extent, it also may suffer from various types of attacks due to the presence of malicious objects [21]. Some malicious objects may launch bad-mouthing or cheating attacks to affect the decision process of service delegation [22]. To address this issue, in recent years, some works in the literature have presented various trust models to solve the problems of trust establishment and relationship maintenance among objects in SIoT [23,24]. Trust is a complex and comprehensive concept in SIoT [25,26]. Specifically, trust not only reflects the security and reliability at the IoT system level, but also reflects the degree of cooperation between two IoT objects when establishing an interactive relationship. The establishment of trust will stimulate cooperation and improve security in the process of service [27–29]. Castelfranchi and Falcone introduced a systematic socio-cognitive trust theory [27]. They proposed a layered model for trust, which consists of five basic ingredients: trustor, trustee, task, goal, and context. They also proposed and analyzed the important characteristics, including integrated, socio-cognitive, multi-factor and multi-dimensional, dynamic, non-prescriptive, etc. The proposed trust theory can be used as a theoretical foundation for analyzing the trust issue of SIoT. Xia et al. combined the fuzzy logic method to solve the trustworthiness convergence issue and proposed a lightweight mechanism for service discovery based on directed acyclic graph (DAG) [28]. On this basis, Xia et al. proposed a trustworthiness inference framework which combines a kernel-based nonlinear multivariate grey prediction model and fuzzy logic method to quantify the trust [29]. Amin et al. presented a classified catalog of friendliness and trust in SIoT. They described the key ingredients and challenges of friendliness- and trust-based approaches, which contributes to the analysis of the effectiveness of the trust model [30]. Narang and Kar proposed a hybrid trust management framework based on probabilistic neighborhood overlap, which considers the resource-constrained IoT devices [31]. Moreover, they analyzed the various attack scenarios, such as slandering/bad-mouthing attack, Sybil attack, self-promoting attack, and ballot stuffing attack to demonstrate the effectiveness of the proposed model. Chen et al. proposed an integrated trust evaluation model which combines direct and indirect trustworthiness [32,33]. Moreover, they further proposed a series of new metrics, such as friendship similarity, social contact similarity, and community of interest similarity to quantify the indirect trust evaluation. They also applied the typical application scenarios, including air pollution detection and augmented map travel assistance, to illustrate the feasibility of the proposed model. In order to comprehensively compare the recent studies along with advantages and disadvantages, we presented detailed comparison of various works in the literature on the SIoT trust model in our previous work [34].

However, the current research on trust model in SIoT still faces three important challenges. First, most works focus on the unidirectional trust evaluation from the SR to the SP. The evaluation of the trustworthiness of SR is ignored, which may cause the trust crisis from the SPs to the SR. The SPs may gradually lose enthusiasm if they suffer prejudiced treatment of the malicious SR. Second, the trust model and service delegation are context- or environment dependent. The properties of the same task are different in different contexts or environments. Third, the decision of service delegation should not only consider the trust of SPs, but also the utility of the SR. In addition, the correlation between trust and utility is ambiguous.

To address the above challenges, we propose a bidirectional trust model and trust-based service delegation approach by comprehensively considering the trust and utility of service requesters and providers. We combine the social trust theory and characteristics of IoT tasks to formalize the trust evaluation and service delegation model. The main contributions of this paper are as follows:

- In order to improve the quality of the IoT service, we propose the bidirectional evaluation and selection model between the SRs and SPs to formulate the process of service or task in SIoT, thus preventing the malicious behaviors of SRs and SPs.

- A context-aware trust model which comprehensively considers the task properties in the specific environment is presented. We employ the subjective logic to construct the opinion-based and evidence-based trust quantification method.
- We present a trust-based service delegation approach that optimizes the utility of the SR while effectively isolating the malicious SPs. Since the service delegation problem in SIoT seldom considers the trust and utility issue at the same time, this paper explores the correlation of trust and utility and their impacts on the service delegation.
- In order to validate the feasibility of our proposed trust model and service delegation method, we present a series of vital experiments to explain the operation of our model. Our results show that the proposed model can effectively assist the IoT object to make the decision of the service delegation.

The remainder of the paper is organized as follows: the system overview and problem statement are presented in Section 2. On this basis, we present the trust and service delegation model, including the trust quantification method and integrated service delegation mechanism in Section 3. In Section 4, in order to demonstrate the feasibility of proposed trust model, we present a series of experiments. In Section 5, we conclude the paper and summarize the contributions. Moreover, some pending research issues are discussed for further research.

2. System Overview and Problem Statement

We consider a general system model in SIoT which consists of five ingredients: (1) service requester (SR), (2) service provider (SP), (3) intermediate object, (4) the context, and the (5) service/task. The life cycle of a task or service is shown in Figure 1. The first step that the SR will perform is to determine the content of the task, including the context, property and goal. It will also publish the task request information. Then, after receiving the message of the task request, the SPs will determine whether to respond to the request by evaluating the trustworthiness of the SR. If the SR is trustworthy from the perspective of the SP, the SP will send a respond message which contains the task price, which is calculated by considering the cost of the task performance. After receiving some response, the SR will delegate the task to the specific SP based on the trust model and the consideration of utility. Then, the delegated SP will perform the task and submit the result. After receiving the result of the task from the delegated SPs, the SR and SP will evaluate each other about their behaviors and update the trust model.

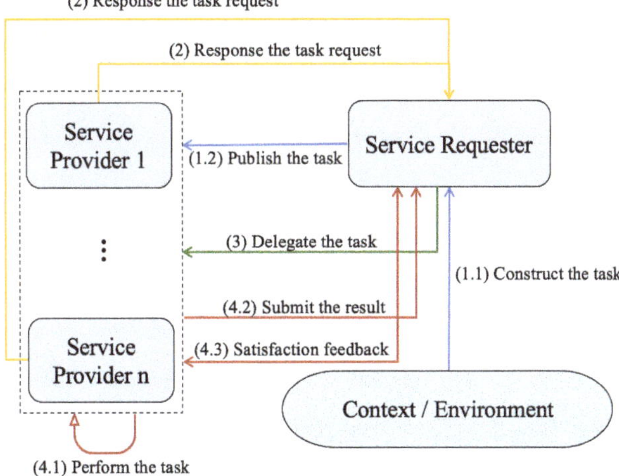

Figure 1. The operation framework of the task/service cycle.

Different from the traditional trust-based service delegation model in SIoT, we combine the bidirectional evaluation to construct the trust model and adopt the utility optimization to formulate the service delegation problem. On the one hand, most of the current literature assumed that SR is reliable, which means the trust between the SR and SP is unidirectional. This assumption may be reasonable and useful in the small-scale network or SR-centric situation. However, in the open and large-scale SIoT scenarios, the SR may not be reliable. If there is no bidirectional evaluation, a malicious SR may damage the SP's privacy, or it may delay a payment after the SP submits the task results. On the other hand, the current literature often employs trustworthiness to determine which SP should be delegated, but there is a lack of consideration for utility issues. To this end, we design the trust-based utility formulation for service delegation.

In order to facilitate the formal description, we divide the entire process into four steps, focusing on the decision-making problem of the object in the process of IoT tasks or services.

2.1. Step 1: The SR Determines the Content of the Task in the Specific Context

In the first step, the SR u_i will comprehensively consider the goal of task and the context to construct the content of the task. A task includes the several necessary properties, which reflect the SR requirements. Formally, the task is denoted by $\varphi = \{p_\varphi = \{p_\varphi^1, p_\varphi^2, \ldots, p_\varphi^m\}|G_\varphi\}_C$, where C is the context of the task φ. p_φ represents the properties of task φ in the context C, and G_φ is the goal of the SR u_i for publishing the task φ.

2.2. Step 2: The SPs Determine Whether to Response the Task Request of the SR

After receiving the request from the SR u_i, the SPs will evaluate the trustworthiness of the SR u_i based on the direct interaction records and some recommendation opinions from several intermediate objects. The set of SPs is denoted by $V = \{v_1, v_2, \ldots, v_n\}$. The set of intermediate objects, which have some interactions with the SR, is denoted by $IN_{SR} = \{r_1, r_2, \ldots\}$. The trustworthiness of the SR u_i from the viewpoint of the SP v_j is formulated as

$$\overrightarrow{T}_{u_i \leftarrow v_j}(\varphi) = f_{ct}(\overrightarrow{T}^d_{u_i \leftarrow v_j}(\varphi), \{\overrightarrow{T}^{rec}_{u_i \leftarrow r_k}(\varphi)\}_{k=1,2,\ldots}), \tag{1}$$

where $\overrightarrow{T}^d_{u_i \leftarrow v_j}(\varphi)$ denotes the direct trust vector of the SR u_i from the viewpoint of the SP v_j. Additionally, $\overrightarrow{T}^{rec}_{u_i \leftarrow r_k}(\varphi)$ denotes the recommendation trust of the SR u_i from the viewpoint of the intermediate object r_k. The function f_{ct} is the convergence function of the trust opinions from the different sources. Based on the evaluation result of the SR trust, the SP will determine whether to respond to the task request by solving the following formulation:

$$Response_{u_i \leftarrow v_j}(\varphi) \begin{cases} \psi_{v_j}(\varphi), & g(\overrightarrow{T}_{u_i \leftarrow v_j}(\varphi)) \geq th_v(\varphi) \\ null, & g(\overrightarrow{T}_{u_i \leftarrow v_j}(\varphi)) < th_v(\varphi). \end{cases} \tag{2}$$

where $th_v(\varphi)$ is a response threshold set by v_j for the task φ, and $\psi_{v_j}(\varphi)$ denotes the price that SR u_i needs to pay to the SP v_j if u_i delegates v_j to perform task φ. $g(\cdot)$ denotes the function of the trustworthiness calculation.

2.3. Step 3: The SR Delegates the Task to the SP

After receiving several responses, the SR u_i will consider the factors of trust and utility to make a decision of service delegation. Similar to the process of trust evaluation of the SR from the viewpoint of the SP in step 2, the trust of the SP v_j from the viewpoint of the SR u_i is formulated as

$$\overrightarrow{T}_{v_j \leftarrow u_i}(\varphi) = f_{ct}(\overrightarrow{T}^d_{v_j \leftarrow u_i}(\varphi), \{\overrightarrow{T}^{rec}_{v_j \leftarrow s_k}(\varphi)\}_{k=1,2,\ldots}), \tag{3}$$

where s_k denotes the intermediate objects which have some interactions with the SP v_j. Based on the trust analysis, the SR will determine the delegated SP by solving the following formulation:

$$DSP = \underset{v_j}{\arg\max}\ f_{tu}(\overrightarrow{T}_{v_j \leftarrow u_i}(\varphi), \psi_{v_j}(\varphi)),$$

$$s.t.\quad g(\overrightarrow{T}_{v_j \leftarrow u_i}(\varphi)) \geq th_u(\varphi)$$

$$(4)$$

where f_{tu} is the delegation function that calculates the integrated index for service delegation. $th_u(\varphi)$ is a trust threshold set by u_i for the task φ.

2.4. Step 4: The Delegated SP Performs the Task and Submits the Result, and Then the SR and SP Will Mutually Comment Each Other

After receiving the delegation message from the SR, the delegated SP (we assume v_j) will perform the task and submit the result. After that, the SR will evaluate the result according to the accuracy, real-time, etc., of the task performance to decide the success or failure of the task. The SR's evaluation of the task is denoted by $Y_{v_j \leftarrow u_i}^{t_\varphi}(\varphi)$, where t_φ is the occurred time of the task φ. If the SR is satisfied according to the SP's performance, the $Y_{v_j \leftarrow u_i}^{t_\varphi}(\varphi)$ will be set 1, and it will be set -1 if the SR is unsatisfied. Similarly, the SP will also evaluate the SR's behavior in the process of the task, which is denoted by $Y_{u_i \leftarrow v_j}^{t_\varphi}(\varphi)$. If the SP is satisfied, then the $Y_{u_i \leftarrow v_j}^{t_\varphi}(\varphi)$ will be set 1, and it will be set -1 if the SP feels unsatisfied.

2.5. Problem Statement

According to the previous description, we can find that in the entire service delegation process, the most important part lies in the rules for mutual trust evaluation between objects, and how to use the trust evaluation information to make decisions. The first important problem is the structuralization of the interactions and the calculation of the direct trust.

Problem Statement 1: *Based on historical interaction records between object A and object B, how does object A determine the direct trust of object B?*

The recommended trust opinions from intermediate objects can be great references for object A to evaluate the trust of object B. However, trust opinions from different sources should have different degrees of confidence. For example, we usually believe in information from reliable sources. Therefore, how to effectively quantify the confidence of information from different sources will be the second important problem.

Problem Statement 2: *When intermediate object C provides A with trust opinions about object B, how will A integrate the opinions of C?*

The success of task execution is seriously related to the delegation decision of the SR, so in the process of service delegation, the SR must carefully evaluate the reliability of SPs. Establishing trust is a suitable way to evaluate the reliability of an object. However, the SR will not only consider trust, but also its own benefits in the delegation process. Therefore, how to comprehensively consider both trust and utility so as to ensure that a relatively reliable SP is selected and optimize the utility of SR is the third important problem.

Problem Statement 3: *According to the trust of the candidate objects, how does object A delegate the task?*

In summary, *Problem 1* corresponds to the quantitative calculation of $T^d_{v_j \leftarrow u_i}(\varphi)$. *Problem 2* corresponds to the formulation of Equation (1). In addition, *Problem 3* corresponds to the formulation of Equation (4).

3. Trust and Service Delegation Model

3.1. Trust Model

In our trust model, the direct interactions and indirect opinions are comprehensively considered. We employ subjective logic for the trust analysis. The results of the trust analysis and utility analysis are integrated for the decision of the service delegation. The whole design framework is shown in Figure 2. Next, we detail the entire trust analysis and service delegation process.

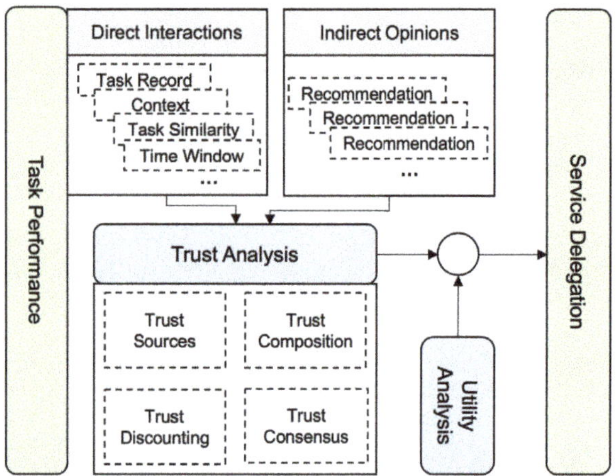

Figure 2. The design framework of the trust model and service delegation.

3.1.1. Subjective Logic

Subjective logic is an uncertain probabilistic logic that was initially introduced by Audun Jøsang to address formal representations of trust [35]. The subjective logic constructs a bijective mapping between opinion space and evidence space, which can help SR to form its own opinion based on the existing direct evidence, and to integrate the recommendation opinions from others to form a comprehensive opinion.

Definition 1 (Opinion Space). *A's direct opinion about object B for the task φ is a vector:*

$$\vec{T}_{B \leftarrow A}(\varphi) = (b_{B \leftarrow A}(\varphi), d_{B \leftarrow A}(\varphi), u_{B \leftarrow A}(\varphi), a_{B \leftarrow A}(\varphi)), \tag{5}$$

where $b_{B \leftarrow A}(\varphi)$ represents the degree to which A believes B will successfully perform the task φ, and $d_{B \leftarrow A}(\varphi)$ represents the degree to which A disbelieves that B will successfully perform the task φ. $u_{B \leftarrow A}(\varphi)$ represents the degree to which A is uncertain about whether B will successfully perform the task φ, and $a_{B \leftarrow A}(\varphi)$ is the base rate. The opinion satisfies the additivity requirement as follows:

$$b_{B \leftarrow A}(\varphi) + d_{B \leftarrow A}(\varphi) + u_{B \leftarrow A}(\varphi) = 1, \tag{6}$$

and the projected probability of the opinion $\vec{T}_{B \leftarrow A}(\varphi)$ is defined as

$$\hat{T}_{B \leftarrow A}(\varphi) = b_{B \leftarrow A}(\varphi) + a_{B \leftarrow A}(\varphi) u_{B \leftarrow A}(\varphi). \tag{7}$$

In our trust model, we use $\overrightarrow{T}^{d}_{B \leftarrow A}(\varphi)$ to represent the direct trust vector of object B from the viewpoint of object A for the task φ. $\hat{T}^{d}_{B \leftarrow A}(\varphi)$ is used for representing the projected trustworthiness of object B.

Evidences are fundamental for forming opinions, which can be presented as a series of the binary comments such as "satisfaction" and "dissatisfaction". The amount of evidence will affect the certainty of the opinion. In subjective logic, the Beta function is used for constructing the evidence space. The probability density function is as follows:

$$Beta(p_x, \alpha, \beta) = \frac{\Gamma(\alpha + \beta)}{\Gamma(\alpha)\Gamma(\beta)} p_x^{\alpha-1}(1 - p_x)^{\beta-1}, \tag{8}$$

where $\Gamma()$ is the gamma function. The beta function can be used to represent the probability distribution of binary events. Therefore, the evidence space can be defined as follows:

Definition 2 (Evidence Space). *An evidence space can be depicted by a beta probability distribution:*

$$Beta(\overrightarrow{T}^{\prime d}_{B \leftarrow A}(\varphi), \alpha_{B \leftarrow A}(\varphi), \beta_{B \leftarrow A}(\varphi)), \tag{9}$$

where $\overrightarrow{T}^{\prime d}_{B \leftarrow A}(\varphi)$ represents the direct trust vector of B from the viewpoint of A in evidence space. $\alpha_{B \leftarrow A}(\varphi)$ and $\beta_{B \leftarrow A}(\varphi)$ are defined as:

$$\begin{cases} \alpha_{B \leftarrow A}(\varphi) = \gamma_{B \leftarrow A}(\varphi) + 2a_{B \leftarrow A}(\varphi) \\ \beta_{B \leftarrow A}(\varphi) = \overline{\gamma}_{B \leftarrow A}(\varphi) + 2(1 - a_{B \leftarrow A}(\varphi)) \end{cases} \tag{10}$$

where $\gamma_{B \leftarrow A}(\varphi)$ and $\overline{\gamma}_{B \leftarrow A}(\varphi)$ are the evidence strength which is based on the historical interactions between objects A and B. $\gamma_{B \leftarrow A}(\varphi)$ denotes the positive evidence strength, which indicates that the B is trustworthy. $\overline{\gamma}_{B \leftarrow A}(\varphi)$ denotes the negative evidence strength, which indicates that B is untrustworthy.

The expected probability $E(\overrightarrow{T}^{\prime d}_{B \leftarrow A}(\varphi))$ is defined as the projected trustworthiness in evidence space, which is expressed as follows:

$$\begin{aligned} \widetilde{T}^{d}_{B \leftarrow A}(\varphi) = E(\overrightarrow{T}^{\prime d}_{B \leftarrow A}(\varphi)) &= \frac{\alpha_{B \leftarrow A}(\varphi)}{\alpha_{B \leftarrow A}(\varphi) + \beta_{B \leftarrow A}(\varphi)} \\ &= \frac{\gamma_{B \leftarrow A}(\varphi) + 2a_{B \leftarrow A}(\varphi)}{\gamma_{B \leftarrow A}(\varphi) + \overline{\gamma}_{B \leftarrow A}(\varphi) + 2} \end{aligned} \tag{11}$$

The bijective mapping between the trust vector in opinion space and the trust vector in the evidence space emerges from the intuitive requirement $\hat{T}^{d}_{B \leftarrow A}(\varphi) = \widetilde{T}^{d}_{B \leftarrow A}(\varphi)$, which is defined as follows.

Definition 3 (Mapping between opinion space and evidence space).

$$\begin{cases} b_{B \leftarrow A}(\varphi) = \frac{\gamma_{B \leftarrow A}(\varphi)}{\gamma_{B \leftarrow A}(\varphi) + \overline{\gamma}_{B \leftarrow A}(\varphi) + 2} \\ d_{B \leftarrow A}(\varphi) = \frac{\overline{\gamma}_{B \leftarrow A}(\varphi)}{\gamma_{B \leftarrow A}(\varphi) + \overline{\gamma}_{B \leftarrow A}(\varphi) + 2} \\ u_{B \leftarrow A}(\varphi) = \frac{2}{\gamma_{B \leftarrow A}(\varphi) + \overline{\gamma}_{B \leftarrow A}(\varphi) + 2} \end{cases} \tag{12}$$

$$\begin{cases} \gamma_{B \leftarrow A}(\varphi) = \frac{2b_{B \leftarrow A}(\varphi)}{u_{B \leftarrow A}(\varphi)} \\ \overline{\gamma}_{B \leftarrow A}(\varphi) = \frac{2d_{B \leftarrow A}(\varphi)}{u_{B \leftarrow A}(\varphi)} \end{cases} \tag{13}$$

3.1.2. Direct Trust

In this paragraph, we introduce the direct trust which is based on the direct interaction records between objects A and B.

- **Task Similarity**

Due to the different context of each task, the importance of different past interaction comments to the current task is different and should be decided by the task similarity. To this end, we use the Jaccard Similarity Index to estimate the similarity of two task in the different context, which is expressed as follows.

$$Sim(\varphi, \varphi') = J(p_\varphi, p_{\varphi'}) = \frac{|p_\varphi \cap p_{\varphi'}|}{|p_\varphi \cup p_{\varphi'}|} \tag{14}$$

For a simple example, we assume there are, in total, four properties, such as {"High Definition", "Least Memory", "Location Range", "Real-Time", and "Measurement Accuracy"}. If the property is required in the task, then the corresponding value of the property vector is set to "1" and otherwise "0". If the φ is a video monitoring task, then the p_φ may be equal to {1, 1, 1, 1, 0}. If the φ' is crowdsensing noise monitoring, then the $p_{\varphi'}$ may be equal to {0, 1, 1, 0, 1}. Then the similarity between φ and φ' is equal to 2/5.

- **Time Window**

The evidence is time dependent. Recent task performance has a greater effect than the older task on the trust evaluation of the object. The time window is presented for the time-dependent strength of single evidence, which is shown in Figure 3.

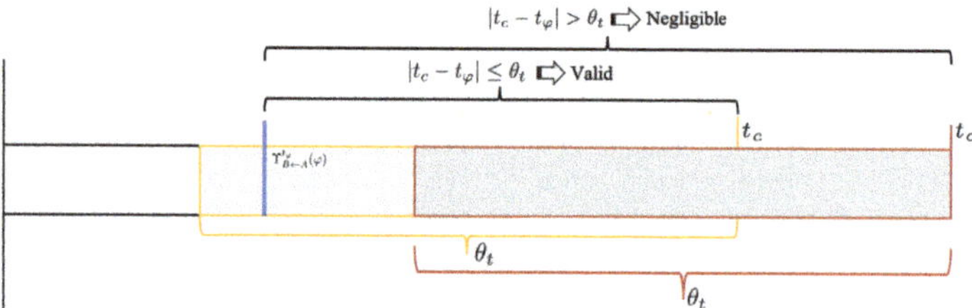

Figure 3. The design of time window for trust evaluation.

Based on the time window, the strength of single evidence can be expressed as follows:

$$\hat{Y}^{t_\varphi}_{B \leftarrow A}(\varphi) = \begin{cases} Y^{t_\varphi}_{B \leftarrow A}(\varphi)e^{-\lambda(t_c - t_\varphi)}, & |t_c - t_\varphi| \leq \theta_t. \\ 0, & |t_c - t_\varphi| > \theta_t. \end{cases} \tag{15}$$

where t_c denotes the current time and λ denotes the decay factor, which affects the rate of decay of the evidence strength.

- **Evidence Strength**

By aggregating the valid direct interaction records, that is, a batch of valid single evidence, we can calculate the total strength of direct evidences as follows:

$$\gamma^d_{B \leftarrow A}(\varphi) = \sum_{\hat{Y}^{t_\varphi}_{B \leftarrow A}(\varphi') > 0} \hat{Y}^{t_\varphi}_{B \leftarrow A}(\varphi')Sim(\varphi, \varphi'), \tag{16}$$

$$\overline{\gamma}^d_{B \leftarrow A}(\varphi) = -\sum_{\hat{Y}^{t_\varphi}_{B \leftarrow A}(\varphi') < 0} \hat{Y}^{t_\varphi}_{B \leftarrow A}(\varphi')Sim(\varphi, \varphi'). \tag{17}$$

- **Direct Trust Calculation**

By combining the methods of task similarity, time window, and evidence strength, we can calculate the direct trust vector $\overrightarrow{T}^d_{B \leftarrow A}(\varphi)$ of the object B from the viewpoint of A for the task φ by substituting Equations (16) and (17) into (12). Therefore, the *problem 1* is addressed through the above design and analysis.

3.1.3. Indirect Trust

In addition to direct trust evaluation, A will also ask C and D for relevant opinions about B. This paragraph solves the fusion problem of recommendation opinions by designing the discounting and consensus operators. The recommendation opinions from objects C and D are expressed as follows, respectively.

$$\overrightarrow{T}^{rec}_{B\leftarrow C}(\varphi) = (b_{B\leftarrow C}(\varphi), d_{B\leftarrow C}(\varphi), u_{B\leftarrow C}(\varphi), a_{B\leftarrow C}(\varphi)) \tag{18}$$

$$\overrightarrow{T}^{rec}_{B\leftarrow D}(\varphi) = (b_{B\leftarrow D}(\varphi), d_{B\leftarrow D}(\varphi), u_{B\leftarrow D}(\varphi), a_{B\leftarrow D}(\varphi)) \tag{19}$$

The objective in this part is to construct the suitable function $f_{ind}(\cdot)$ to integrate the $\overrightarrow{T}^{rec}_{B\leftarrow D}(\varphi)$ and $\overrightarrow{T}^{rec}_{B\leftarrow C}(\varphi)$, which is formulated as follows:

$$\overrightarrow{T}^{ind}_{B\leftarrow A}(\varphi) = f_{ind}(\overrightarrow{T}^{rec}_{B\leftarrow C}(\varphi), \overrightarrow{T}^{rec}_{B\leftarrow D}(\varphi)), \tag{20}$$

and $\overrightarrow{T}^{ind}_{B\leftarrow A}(\varphi) = (b^{ind}_{B\leftarrow A}, d^{ind}_{B\leftarrow A}, u^{ind}_{B\leftarrow A}, a^{ind}_{B\leftarrow A})$.

- **Discounting and Consensus Operator**

In the subjective logic framework, the discounting rule does not have a natural interpretation of evidence handling [36]. To this end, we use the trust in opinion space to discount the trust in evidence space. The ideas and principles are shown in Figure 4. We use the symbol $\otimes$ to represent the discounting operator. Thus we have $\overrightarrow{T}^{ind}_{B\leftarrow C\leftarrow A}(\varphi) = \overrightarrow{T}^{rec}_{B\leftarrow C}(\varphi) \otimes \overrightarrow{T}^{d}_{C\leftarrow A}(\varphi)$.

The specific discounting rule $\otimes$ in evidence space is shown as follows.

$$\begin{cases} \gamma^{ind}_{B\leftarrow C\leftarrow A}(\varphi) = \hat{T}^{d}_{C\leftarrow A}(\varphi)\gamma^{rec}_{B\leftarrow C}(\varphi) \\ \overline{\gamma}^{ind}_{B\leftarrow C\leftarrow A}(\varphi) = \hat{T}^{d}_{C\leftarrow A}(\varphi)\overline{\gamma}^{rec}_{B\leftarrow C}(\varphi) \end{cases} \tag{21}$$

Based on Equations (12) and (21), we can calculate the indirect trust vector $\overrightarrow{T}^{ind}_{B\leftarrow C\leftarrow A}(\varphi)$, which is shown as follows.

$$\begin{cases} b^{ind}_{B\leftarrow C\leftarrow A}(\varphi) = \dfrac{\hat{T}^{d}_{C\leftarrow A}(\varphi)b^{rec}_{B\leftarrow C}(\varphi)}{\hat{T}^{d}_{C\leftarrow A}(\varphi)b^{rec}_{B\leftarrow C}(\varphi) + \hat{T}^{d}_{C\leftarrow A}(\varphi)d^{rec}_{B\leftarrow C}(\varphi) + u^{rec}_{B\leftarrow C}(\varphi)} \\[3mm] d^{ind}_{B\leftarrow C\leftarrow A}(\varphi) = \dfrac{\hat{T}^{d}_{C\leftarrow A}(\varphi)d^{rec}_{B\leftarrow C}(\varphi)}{\hat{T}^{d}_{C\leftarrow A}(\varphi)b^{rec}_{B\leftarrow C}(\varphi) + \hat{T}^{d}_{C\leftarrow A}(\varphi)d^{rec}_{B\leftarrow C}(\varphi) + u^{rec}_{B\leftarrow C}(\varphi)} \\[3mm] u^{ind}_{B\leftarrow C\leftarrow A}(\varphi) = \dfrac{u^{rec}_{B\leftarrow C}(\varphi)}{\hat{T}^{d}_{C\leftarrow A}(\varphi)b^{rec}_{B\leftarrow C}(\varphi) + \hat{T}^{d}_{C\leftarrow A}(\varphi)d^{rec}_{B\leftarrow C}(\varphi) + u^{rec}_{B\leftarrow C}(\varphi)} \\[3mm] a^{ind}_{B\leftarrow C\leftarrow A}(\varphi) = \hat{T}^{d}_{C\leftarrow A}(\varphi)a^{rec}_{B\leftarrow A}(\varphi) \end{cases} \tag{22}$$

The consensus operator is designed for integrating the recommendation opinions from different sources. We use the weighted sum method to design the consensus operator. Similar to the design idea of discounting operator, we use the trust in opinion space as weight parameters. The symbol $\oplus$ represents the consensus operator, and thus we have $\overrightarrow{T}^{ind}_{B\leftarrow A}(\varphi) = \overrightarrow{T}^{ind}_{B\leftarrow CD\leftarrow A}(\varphi) = \overrightarrow{T}^{ind}_{B\leftarrow C\leftarrow A}(\varphi) \oplus \overrightarrow{T}^{ind}_{B\leftarrow D\leftarrow A}(\varphi)$. The specific consensus operator in evidence space is shown as follows.

$$\begin{cases} \gamma^{ind}_{B\leftarrow A}(\varphi) = \dfrac{(1-u^{ind}_{B\leftarrow C\leftarrow A}(\varphi))\gamma^{ind}_{B\leftarrow C\leftarrow A}(\varphi) + (1-u^{ind}_{B\leftarrow D\leftarrow A}(\varphi))\gamma^{ind}_{B\leftarrow D\leftarrow A}(\varphi)}{(1-u^{ind}_{B\leftarrow C\leftarrow A}(\varphi)) + (1-u^{ind}_{B\leftarrow D\leftarrow A}(\varphi))} \\[3mm] \overline{\gamma}^{ind}_{B\leftarrow A}(\varphi) = \dfrac{(1-u^{ind}_{B\leftarrow C\leftarrow A}(\varphi))\overline{\gamma}^{ind}_{B\leftarrow C\leftarrow A}(\varphi) + (1-u^{ind}_{B\leftarrow D\leftarrow A}(\varphi))\overline{\gamma}^{ind}_{B\leftarrow D\leftarrow A}(\varphi)}{(1-u^{ind}_{B\leftarrow C\leftarrow A}(\varphi)) + (1-u^{ind}_{B\leftarrow D\leftarrow A}(\varphi))} \end{cases} \tag{23}$$

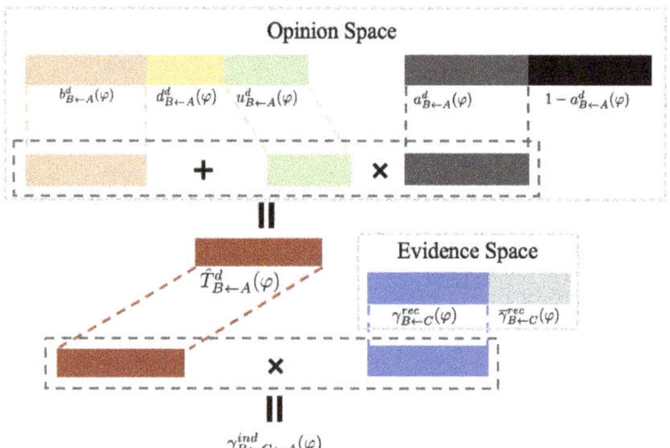

Figure 4. The design of discounting operator.

- **Indirect Trust Calculation**

  From Equations (12) and (23), we obtain the indirect trust vector $\overrightarrow{T}^{ind}_{B \leftarrow A}(\varphi)$:

$$
\begin{cases}
b^{ind}_{B \leftarrow A}(\varphi) = \dfrac{(1 - u^{ind}_{B \leftarrow C \leftarrow A}(\varphi)) u^{ind}_{B \leftarrow D \leftarrow A}(\varphi) b^{ind}_{B \leftarrow C \leftarrow A}(\varphi) + (1 - u^{ind}_{B \leftarrow D \leftarrow A}(\varphi)))^{ind}_{B \leftarrow C \leftarrow A}(\varphi) b^{ind}_{B \leftarrow D \leftarrow A}(\varphi)}{(1 - u^{ind}_{B \leftarrow C \leftarrow A}(\varphi)) u^{ind}_{B \leftarrow D \leftarrow A}(\varphi) + (1 - u^{ind}_{B \leftarrow D \leftarrow A}(\varphi)) u^{ind}_{B \leftarrow C \leftarrow A}(\varphi)} \\[1.2em]
d^{ind}_{B \leftarrow A}(\varphi) = \dfrac{(1 - u^{ind}_{B \leftarrow C \leftarrow A}(\varphi)) u^{ind}_{B \leftarrow D \leftarrow A}(\varphi) d^{ind}_{B \leftarrow C \leftarrow A}(\varphi) + (1 - u^{ind}_{B \leftarrow D \leftarrow A}(\varphi)) u^{ind}_{B \leftarrow C \leftarrow A}(\varphi) d^{ind}_{B \leftarrow D \leftarrow A}(\varphi)}{(1 - u^{ind}_{B \leftarrow C \leftarrow A}(\varphi)) u^{ind}_{B \leftarrow D \leftarrow A}(\varphi) + (1 - u^{ind}_{B \leftarrow D \leftarrow A})(\varphi) u^{ind}_{B \leftarrow C \leftarrow A}(\varphi)} \\[1.2em]
u^{ind}_{B \leftarrow A}(\varphi) = \dfrac{(1 - u^{ind}_{B \leftarrow C \leftarrow A}(\varphi)) u^{ind}_{B \leftarrow C \leftarrow A}(\varphi) u^{ind}_{B \leftarrow D \leftarrow A}(\varphi) + (1 - u^{ind}_{B \leftarrow D \leftarrow A}(\varphi)) u^{ind}_{B \leftarrow C \leftarrow A}(\varphi) u^{ind}_{B \leftarrow D \leftarrow A}(\varphi)}{(1 - u^{ind}_{B \leftarrow C \leftarrow A}(\varphi)) u^{ind}_{B \leftarrow D \leftarrow A}(\varphi) + (1 - u^{ind}_{B \leftarrow D \leftarrow A}(\varphi)) u^{ind}_{B \leftarrow C \leftarrow A}(\varphi)} \\[1.2em]
a^{ind}_{B \leftarrow A}(\varphi) = \dfrac{(1 - u^{ind}_{B \leftarrow C \leftarrow A}(\varphi)) a^{ind}_{B \leftarrow C \leftarrow A}(\varphi) + (1 - u^{ind}_{B \leftarrow D \leftarrow A}(\varphi)) a^{ind}_{B \leftarrow D \leftarrow A}(\varphi)}{(1 - u^{ind}_{B \leftarrow C \leftarrow A}(\varphi)) + (1 - u^{ind}_{B \leftarrow D \leftarrow A}(\varphi))}
\end{cases}
\tag{24}
$$

Therefore, we have the indirect trust calculation function

$$
\begin{aligned}
f_{ind}(\overrightarrow{T}^{rec}_{B \leftarrow C}(\varphi), \overrightarrow{T}^{rec}_{B \leftarrow D}(\varphi)) = \\
(\overrightarrow{T}^{rec}_{B \leftarrow C}(\varphi) \otimes \overrightarrow{T}^{d}_{C \leftarrow A}(\varphi)) \oplus (\overrightarrow{T}^{rec}_{B \leftarrow D}(\varphi) \otimes \overrightarrow{T}^{d}_{D \leftarrow A}(\varphi)).
\end{aligned}
\tag{25}
$$

3.1.4. Compositive Trust

The compositive trust is the fusion of direct trust and indirect trust. We also use the consensus operator to fuse them. From Equations (3) and (25), we have

$$
\begin{aligned}
\overrightarrow{T}_{B \leftarrow A}(\varphi) &= f_{ct}(\overrightarrow{T}^{d}_{B \leftarrow A}(\varphi), \overrightarrow{T}^{rec}_{B \leftarrow C}(\varphi), \overrightarrow{T}^{rec}_{B \leftarrow D}(\varphi)) \\
&= \overrightarrow{T}^{d}_{B \leftarrow A}(\varphi) \oplus \overrightarrow{T}^{ind}_{B \leftarrow A}(\varphi) \\
&= \overrightarrow{T}^{d}_{B \leftarrow A}(\varphi) \oplus [(\overrightarrow{T}^{rec}_{B \leftarrow C}(\varphi) \otimes \overrightarrow{T}^{d}_{C \leftarrow A}(\varphi)) \oplus (\overrightarrow{T}^{rec}_{B \leftarrow D}(\varphi) \otimes \overrightarrow{T}^{d}_{D \leftarrow A}(\varphi))].
\end{aligned}
\tag{26}
$$

Therefore, *problem 2* is addressed through the above design and analysis.

3.2. Service Delegation Mechanism

After calculating the trust vector of the SP v_j based on the method proposed at last subsection, we further study the issue of service delegation. In SIoT, the SR u_i will not only consider the trust of the SP v_j, but also concern the utility. Therefore, we present the trust-based service delegation method to solve the *Problem 3*. We define the decision function of service delegation as follows:

$$
\begin{aligned}
U_{v_j \leftarrow u_i}(\varphi) &= f_{tu}(\overrightarrow{T}_{v_j \leftarrow u_i}(\varphi), \psi_{v_j}(\varphi)) \\
&= \hat{T}_{v_j \leftarrow u_i}(\varphi)(\zeta_{u_i}(\varphi) - \psi_{v_j}(\varphi)) + (d_{v_j \leftarrow u_i}(\varphi) + (1 - a_{v_j \leftarrow u_i}(\varphi)) u_{v_j \leftarrow u_i})(-\bar{\zeta}_{u_i}(\varphi)),
\end{aligned}
\tag{27}
$$

where $\psi_{v_j}(\varphi)$ denotes the benefit value when the task is successful and $\overline{\zeta}_{u_i}(\varphi)$ denotes the lost value when the task is failed. Therefore, the decision of the service delegation (e.g., Equation (4)) can be rewritten as follows:

$$DSP = \arg\max_{v_j}\ U_{v_j \leftarrow u_i}(\varphi)$$
$$s.t.\quad \hat{T}_{v_j \leftarrow u_i}(\varphi)) \geq th_u(\varphi) \tag{28}$$

Through the proposed decision-making method for service delegation, the SR can make a plan to maximize its own utility under the consideration of trust of SPs. For the entire SIoT system, on the one hand, our proposed method can guarantee a high task success rate based on trust analysis. On the other hand, we can improve the overall social welfare and boost the cooperation.

4. Simulation and Results

In order to verify the validity of the subjective logic-based trust model proposed in this section, this study conducts experiments based on the Netlogo experimental platform [37]. NetLogo is an agent-based programming language, which is useful to simulate the interaction among objects and monitor the state changes in a simulative SIoT environment. The construction of the experimental platform is based on our previous work [38]. The trust evaluation mechanism module and service delegation module are adjusted based on the aforementioned bidirectional model. The experiments are divided into the following parts: First, after the interactive experiment, the results of the bidirectional trust evaluation of SPs to SR and SR to SPs are observed to test the effectiveness of subjective logic in the process of trust evaluation. On this basis, the impact of similarity of the services/tasks and positive evaluation rates on trust evaluation results are analyzed. Then, the influence of the number of recommenders on the compositive trust evaluation results is analyzed, and finally the benefits of SR and the changes in the number of responding SPs are measured.

This study defines the rate of positive evidence (RPE) as the proportion of the number of simulated service results that are rated as "positive—that is, satisfied" in the total number of service evaluations. Similarity, the rate of task similarity (RTS) is the similarity of the attributes among the services. For example, when the similarity is 40%, it means that 40% attributes of randomly generated services in the network are consistent. In this experiment, a total of 110 virtual nodes are deployed for service interaction, of which 10 nodes are employed as SRs and 100 nodes are employed as SPs. At the same time, the above virtual nodes will also serve as intermediate nodes in the process of trust evaluation to provide recommendations.

4.1. Comparison of SR and SPs' Basic Bidirectional Trust Evaluation Results

In this part of the experiment, the positive evaluation rate is set to 50%, and the task similarity is 40%. The experiment runs 500 ticks, and one service/task is executed in each tick. In addition, 10 SPs and 1 SR were randomly selected for observation. Figure 5a,b shows the trust evaluation results of 10 SPs and SR. As shown in Figure 5a, compared with the direct trust evaluation results of each SP, the compositive trust evaluation results for SR have less difference and more comprehensive opinions, which reflects that the evaluation method based on subjective logic can better integrate the recommendations from different sources so that most SPs can have a more consistent evaluation for SR. Similarly, as shown in Figure 5b, after the SR obtains the recommendations of other intermediate nodes in the network, it obtains the integrated evaluation results of each SP's trust. It can be seen that the recommendations of other intermediate nodes will facilitate the SR to make a more accurate evaluation on the trustworthiness of SPs.

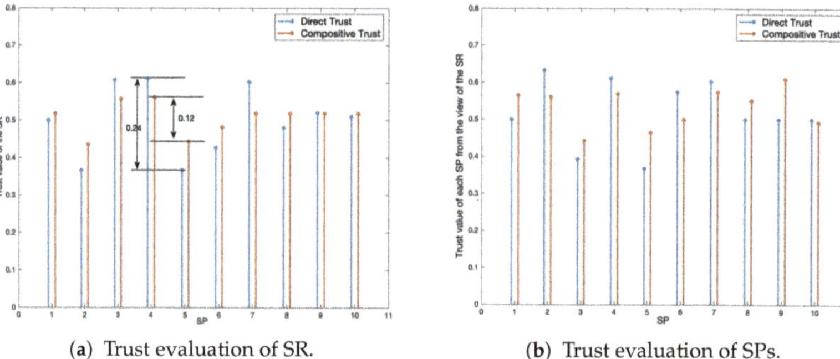

(a) Trust evaluation of SR. (b) Trust evaluation of SPs.

Figure 5. Bidirectional trust evaluation of SR and SPs.

4.2. The Influence of RPE and RTS on the Results of Trust Evaluation

This part of the experiment analyzes the impact of RPE and RTS on the evaluation of SR's trust. As shown in Figure 6a, with the increase in the positive evaluation rate, the trust evaluation result of the object will be improved to a certain extent. However, this improvement still has certain limitations. The evaluation results of some SPs for SR may decrease with the increase in RPE. The main reason is that due to the low similarity of tasks. Although some service evaluation opinions are positive or satisfactory, the evidence strength is slight.

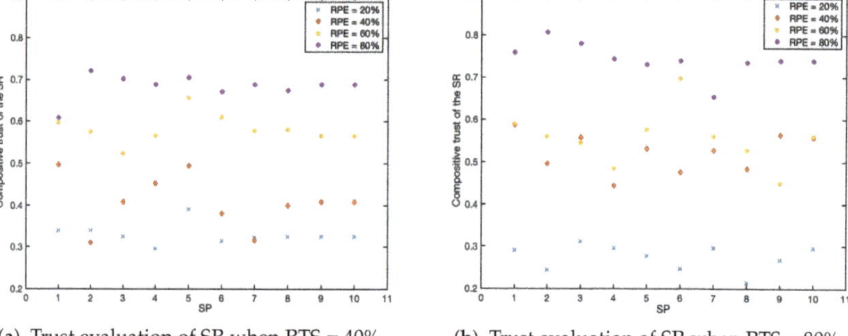

(a) Trust evaluation of SR when RTS = 40%. (b) Trust evaluation of SR when RTS = 80%.

Figure 6. Trust evaluation of SR with different RPE and RTS.

As shown in Figure 6b, compared with the case where the task similarity is 40%, when the task similarity is 80%, the attributes between tasks are more similar. Therefore, the evidence strength of a single evidence will be increased, which will make the formation of the trust evaluation more accurate and reliable. Compared with Figure 6a, the upper and lower boundaries in Figure 6b are larger, and the differences among different RPE groups are more obvious. It can be demonstrated that when the task similarity is greater, the object can provide more accurate recommendations, thereby forming a more accurate trust evaluation result.

4.3. The Influence of the Number of Recommenders on the Trust Evaluation

In the process of trust evaluation for a certain SP, the SR needs to collect the recommendation opinions from the intermediate nodes to form a more accurate trust point of view. As shown in Figure 7, when the number of the recommenders is 0, it indicates that the trust value of SP to form the viewpoint of the SR is completely evaluated based on direct experience. Along with the number of recommenders gradually increasing, the SR can collect more recommendation opinions. From the experimental results of this group, it can be seen that when the number of recommendation opinions is equal to or greater than 6, the SR's trust evaluation opinion on SP tends to be stable, and the SR can more accurately identify the honest and trustworthy SP while avoiding wrongly delegating malicious or negative SPs. Therefore, in order to better evaluate the trustworthiness of the SP, the SR needs to obtain as many recommendations from intermediate nodes as possible during the service delegation process.

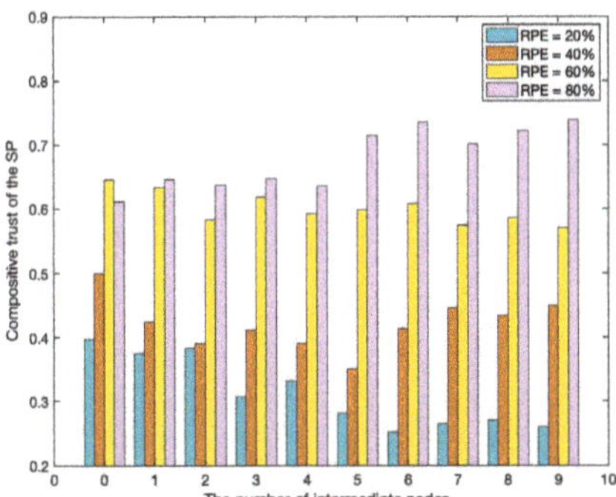

Figure 7. Trust evaluation of the SP with different number of intermediate nodes.

4.4. Quantity of Responsive SPs and Benefit Analysis of the SR under the Bidirectional Trust Evaluation

Figure 8 shows the number of responsive SPs and benefits of the SR when the RPE is from 10% to 90% for a certain task. It can be seen that when the RPE is less than 0.5 and the RTS is large, there are more negative opinions referenced. Therefore, the trust evaluation result of SR from the viewpoint of SPs is generally low, and few SPs respond. Therefore, the SR cannot select a suitable SP, and the income is low. With the increase in RPE, the trustworthiness of the SR increases and the number of responding SPs gradually increases, so the SR can obtain a better delegation scheme, which improves the overall revenue. In addition, in the case of small RPE, although the expected benefit of SR is higher when the RTS is lower (the reason is that some SPs cannot correctly estimate the trustworthiness of SR, resulting in a wrong response to the service), it may lead to lower service quality of SPs and failure to guarantee the benefits of SPs. Higher RTS will lead to more accurate bidirectional evaluation results, and more SPs choose not to respond to the service request when RPE is low. On the other hand, in the case of higher RPE, higher RTS will make the bidirectional evaluation between SPs and SR more accurate, so the overall benefit of the SR will be higher.

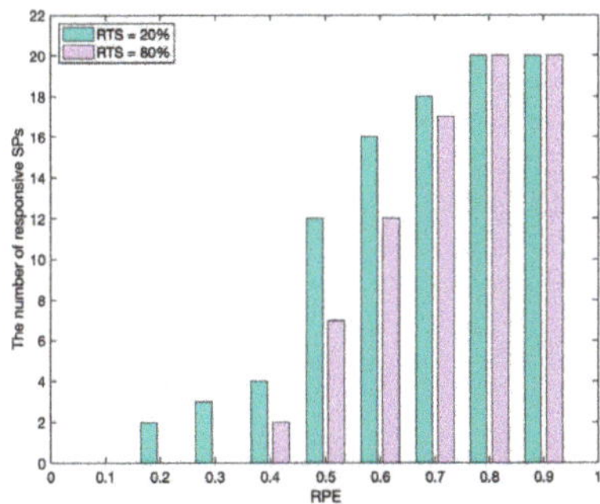

(**a**) The number of responsive SPs with different RPE.

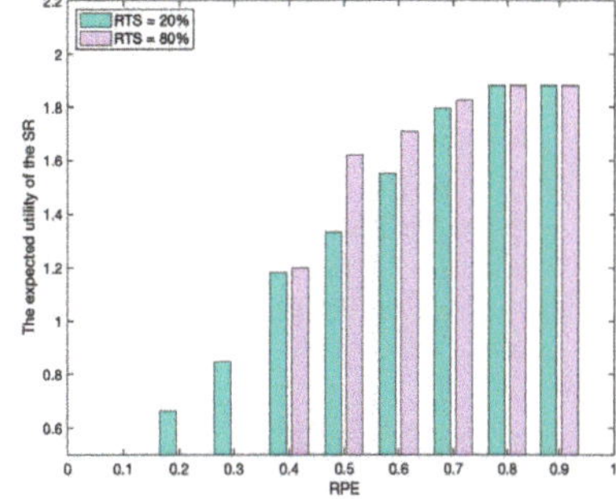

(**b**) Expected utility of the SR with different RPE.

Figure 8. The number of responsive SPs and the SR's utility with different RPE.

5. Conclusions and Discussions

In this article, we studied the trust-based service delegation problem in SIoT. Considering the bidirectionality of trust, we design a framework of the trust model and service delegation. On this basis, we propose a bidirectional trust evaluation method based on subjective logic. We have shown that by using this formulation, the SR and SP can quantitatively evaluate the trust of each other in a reasonable way. In addition, we consider the context of the task to ensure the feasibility of our model in the SIoT scenario. The task similarity and time window are presented for the calculation of evidence strength. The convergence operators including discounting and consensus operator are constructed for compositive trust quantification. The decision-making approach of the service delegation with comprehensive consideration of trust and utility is proposed to ensure the success of the task while improving the utility of the SR.

However, the current work is in infancy. First, considering the computational complexity, the proposed model simplifies the condition setting to a certain extent. The evidence composition in evidence space only includes service attributes, bidirectional service evaluation information, service time, etc., without considering the relationships between device characteristics of IoT objects and service properties. Therefore, our proposed model is more suitable for the scenarios where the degree of heterogeneity and differentiation of IoT devices is low. The evidence-based descriptions of the characteristics of IoT devices and the relationship between these evidence-based descriptions and opinions will be our important future work. Second, with the development of the Internet of Things, some new architectures, such as multiple internets of things, are proposed. Therefore, we will further evaluate whether our model can be feasible and adaptive for various paradigms [39–41]. Moreover, we plan to extend this model and configure a real-world application scenario in order to make it more capable. The task simulations at different network scales will be carried out in the following research process to validate the effectiveness and practicability of our trust model and service delegation method. Furthermore, testing under different attack environments will be also further provided.

Author Contributions: Conceptualization, L.W., C.L. and Y.-B.L.; Methodology, L.W. and J.W.; Software, L.W. and Y.Y.; Validation, L.W., Y.Y., J.W., C.L. and Y.-B.L.; Formal Analysis, L.W. and C.L.; Investigation, L.W., Y.Y. and J.W.; Writing—original Draft Preparation, L.W., Y.Y. and Y.-B.L.; Writing—review and Editing, L.W., J.W. and Y.-B.L.; Supervision, C.L. All authors have read and agreed to the published version of the manuscript.

Funding: This work was supported by the National Natural Science Foundation of China under Grants 62136006, 62073215 and 61873166.

Data Availability Statement: Not applicable.

Conflicts of Interest: The authors declare no conflict of interest.

References

1. Atzori, L.; Iera, A.; Morabito, G. The Internet of Things: A survey. *Comput. Netw.* **2010**, *54*, 2787–2805. [CrossRef]
2. Li, S.; Da Xu, L.; Zhao, S. The internet of things: A survey. *Inf. Syst. Front.* **2015**, *17*, 243–259. [CrossRef]
3. Tan, L.; Wang, N. Future internet: The internet of things. In Proceedings of the 2010 3rd International Conference on Advanced Computer Theory and Engineering(ICACTE), Chengdu, China, 20–22 August 2010; Volume 5, pp. 376–380.
4. Chen, T.; Barbarossa, S.; Wang, X.; Giannakis, G.B.; Zhang, Z.L. Learning and Management for Internet of Things: Accounting for Adaptivity and Scalability. *Proc. IEEE* **2019**, *107*, 778–796. [CrossRef]
5. Silva, J.d.C.; Rodrigues, J.J.P.C.; Al-Muhtadi, J.; Rabêlo, R.A.L.; Furtado, V. Management Platforms and Protocols for Internet of Things: A Survey. *Sensors* **2019**, *19*, 676. [CrossRef]
6. Atzori, L.; Iera, A.; Morabito, G. SIoT: Giving a Social Structure to the Internet of Things. *IEEE Commun. Lett.* **2011**, *15*, 1193–1195. [CrossRef]
7. Atzori, L.; Iera, A.; Morabito, G.; Nitti, M. The social internet of things (siot)–when social networks meet the internet of things: Concept, architecture and network characterization. *Comput. Netw.* **2012**, *56*, 3594–3608. [CrossRef]
8. Vegni, A.M.; Loscrí, V. A Survey on Vehicular Social Networks. *IEEE Commun. Surv. Tutor.* **2015**, *17*, 2397–2419. [CrossRef]
9. Jain, B.; Brar, G.; Malhotra, J.; Rani, S.; Ahmed, S.H. A cross layer protocol for traffic management in Social Internet of Vehicles. *Future Gener. Comput. Syst.* **2018**, *82*, 707–714. [CrossRef]

10. Zia, K.; Shafi, M.; Farooq, U. Improving Recommendation Accuracy Using Social Network of Owners in Social Internet of Vehicles. *Future Internet* **2020**, *12*, 69. [CrossRef]
11. Schurgot, M.R.; Comaniciu, C.; Jaffres-Runser, K. Beyond traditional DTN routing: Social networks for opportunistic communication. *IEEE Commun. Mag.* **2012**, *50*, 155–162. [CrossRef]
12. Wang, J.; Wang, F.; Wang, Y.; Zhang, D.; Wang, L.; Qiu, Z. Social-Network-Assisted Worker Recruitment in Mobile Crowd Sensing. *IEEE Trans. Mob. Comput.* **2019**, *18*, 1661–1673. [CrossRef]
13. Chen, P.Y.; Cheng, S.M.; Ting, P.S.; Lien, C.W.; Chu, F.J. When crowdsourcing meets mobile sensing: A social network perspective. *IEEE Commun. Mag.* **2015**, *53*, 157–163. [CrossRef]
14. Nie, J.; Luo, J.; Xiong, Z.; Niyato, D.; Wang, P. A Stackelberg Game Approach Toward Socially-Aware Incentive Mechanisms for Mobile Crowdsensing. *IEEE Trans. Wirel. Commun.* **2019**, *18*, 724–738. [CrossRef]
15. Wei, L.; Wu, J.; Long, C. A Blockchain-Based Hybrid Incentive Model for Crowdsensing. *Electronics* **2020**, *9*, 215. [CrossRef]
16. Hu, X.; Li, X.; Ngai, E.C.H.; Leung, V.C.; Kruchten, P. Multidimensional context-aware social network architecture for mobile crowdsensing. *IEEE Commun. Mag.* **2014**, *52*, 78–87. [CrossRef]
17. Manogaran, G.; Rodrigues, J.J.P.C.; Kozlov, S.A.; Manokaran, K. Conditional Support-Vector-Machine-Based Shared Adaptive Computing Model for Smart City Traffic Management. *IEEE Trans. Comput. Soc. Syst.* **2022**, *9*, 174–183. [CrossRef]
18. Amin, F.; Choi, G.S. Hotspots Analysis Using Cyber-Physical-Social System for a Smart City. *IEEE Access* **2020**, *8*, 122197–122209. [CrossRef]
19. Azeroual, O.; Jha, M.; Nikiforova, A.; Sha, K.; Alsmirat, M.; Jha, S. A Record Linkage-Based Data Deduplication Framework with DataCleaner Extension. *Multimodal Technol. Interact.* **2022**, *6*, 27. [CrossRef]
20. Rehman, A.U.; Naqvi, R.A.; Rehman, A.; Paul, A.; Sadiq, M.T.; Hussain, D. A Trustworthy SIoT Aware Mechanism as an Enabler for Citizen Services in Smart Cities. *Electronics* **2020**, *9*, 918. [CrossRef]
21. Huang, Z.; Zeng, D.; Chen, H. A comparison of collaborative-filtering recommendation algorithms for e-commerce. *IEEE Intell. Syst.* **2007**, *22*, 68–78. [CrossRef]
22. Guo, J.; Chen, R.; Tsai, J.J. A survey of trust computation models for service management in internet of things systems. *Comput. Commun.* **2017**, *97*, 1–14. [CrossRef]
23. Chahal, R.K.; Kumar, N.; Batra, S. Trust management in social Internet of Things: A taxonomy, open issues, and challenges. *Comput. Commun.* **2020**, *150*, 13–46. [CrossRef]
24. Khan, W.Z.; Arshad, Q.u.A.; Hakak, S.; Khan, M.K.; Saeed-Ur-Rehman. Trust Management in Social Internet of Things: Architectures, Recent Advancements, and Future Challenges. *IEEE Internet Things J.* **2021**, *8*, 7768–7788. [CrossRef]
25. Roopa, M.S.; Pattar, S.; Buyya, R.; Venugopal, K.R.; Iyengar, S.S.; Patnaik, L.M. Social Internet of Things (SIoT): Foundations, thrust areas, systematic review and future directions. *Comput. Commun.* **2019**, *139*, 32–57. [CrossRef]
26. Nitti, M.; Girau, R.; Atzori, L. Trustworthiness Management in the Social Internet of Things. *IEEE Trans. Knowl. Data Eng.* **2014**, *26*, 1253–1266. [CrossRef]
27. Castelfranchi, C.; Falcone, R. *Trust Theory: A Socio-Cognitive and Computational Model*; John Wiley & Sons: Hoboken, NJ, USA, 2010; Volume 18.
28. Xia, H.; Xiao, F.; Zhang, S.S.; Cheng, X.G.; Pan, Z.K. A reputation-based model for trust evaluation in social cyber-physical systems. *IEEE Trans. Netw. Sci. Eng.* **2018**, *7*, 792–804. [CrossRef]
29. Xia, H.; Xiao, F.; Zhang, S.S.; Hu, C.Q.; Cheng, X.Z. Trustworthiness inference framework in the social Internet of Things: A context-aware approach. In Proceedings of the IEEE INFOCOM 2019-IEEE Conference on Computer Communications, Paris, France, 29 April–2 May 2019; pp. 838–846.
30. Amin, F.; Ahmad, A.; Sang Choi, G. Towards Trust and Friendliness Approaches in the Social Internet of Things. *Appl. Sci.* **2019**, *9*, 166. [CrossRef]
31. Narang, N.; Kar, S. A hybrid trust management framework for a multi-service social IoT network. *Comput. Commun.* **2021**, *171*, 61–79. [CrossRef]
32. Chen, R.; Guo, J.; Bao, F. Trust management for SOA-based IoT and its application to service composition. *IEEE Trans. Serv. Comput.* **2014**, *9*, 482–495. [CrossRef]
33. Chen, R.; Bao, F.; Guo, J. Trust-based service management for social internet of things systems. *IEEE Trans. Dependable Secur. Comput.* **2015**, *13*, 684–696. [CrossRef]
34. Wei, L.; Yang, Y.; Wu, J.; Long, C.; Li, B. Trust Management for Internet of Things: A Comprehensive Study. *IEEE Internet Things J.* 2021, Early Access. [CrossRef]
35. Jøsang, A. *Subjective Logic: A Formalism for Reasoning Under Uncertainty*; Springer: Cham, Switzerland, 2016.
36. Škorić, B.; de Hoogh, S.J.A.; Zannone, N. Flow-based reputation with uncertainty: Evidence-based subjective logic. *Int. J. Inf. Secur.* **2016**, *15*, 381–402. [CrossRef]
37. Wilensky, U. *NetLogo. Center for Connected Learning and Computer-Based Modeling*; Northwestern University: Evanston, IL, USA, 1999. Available online: http://ccl.northwestern.edu/netlogo/ (accessed on 16 February 2022).
38. Wei, L.; Wu, J.; Long, C.; Li, B. On Designing Context-Aware Trust Model and Service Delegation for Social Internet of Things. *IEEE Internet Things J.* **2021**, *8*, 4775–4787. [CrossRef]
39. Baldassarre, G.; Lo Giudice, P.; Musarella, L.; Ursino, D. The MIoT paradigm: Main features and an "ad-hoc" crawler. *Future Gener. Comput. Syst.* **2019**, *92*, 29–42. [CrossRef]

40. Cauteruccio, F.; Cinelli, L.; Fortino, G.; Savaglio, C.; Terracina, G.; Ursino, D.; Virgili, L. An approach to compute the scope of a social object in a Multi-IoT scenario. *Pervasive Mob. Comput.* **2020**, *67*, 101223. [CrossRef]
41. Ursino, D.; Virgili, L. An approach to evaluate trust and reputation of things in a Multi-IoTs scenario. *Computing* **2020**, *102*, 2257–2298. [CrossRef]

future internet

MDPI

Article

Query Processing in Blockchain Systems: Current State and Future Challenges

Dennis Przytarski [1,*], Christoph Stach [1], Clémentine Gritti [2] and Bernhard Mitschang [1]

[1] Institute for Parallel and Distributed Systems, University of Stuttgart, Universitätsstraße 38, 70569 Stuttgart, Germany; Christoph.Stach@ipvs.uni-stuttgart.de (C.S.); Bernhard.Mitschang@ipvs.uni-stuttgart.de (B.M.)
[2] Department of Computer Science and Software Engineering, University of Canterbury, Christchurch 8041, New Zealand; clementine.gritti@canterbury.ac.nz
* Correspondence: dennis.przytarski@ipvs.uni-stuttgart.de; Tel.: +49-711-68588-235

Abstract: When, in 2008, Satoshi Nakamoto envisioned the first distributed database management system that relied on cryptographically secured chain of blocks to store data in an immutable and tamper-resistant manner, his primary use case was the introduction of a digital currency. Owing to this use case, the blockchain system was geared towards efficient storage of data, whereas the processing of complex queries, such as provenance analyses of data history, is out of focus. The increasing use of Internet of Things technologies and the resulting digitization in many domains, however, have led to a plethora of novel use cases for a secure digital ledger. For instance, in the healthcare sector, blockchain systems are used for the secure storage and sharing of electronic health records, while the food industry applies such systems to enable a reliable food-chain traceability, e.g., to prove compliance with cold chains. In these application domains, however, querying the current state is not sufficient—comprehensive history queries are required instead. Due to these altered usage modes involving more complex query types, it is questionable whether today's blockchain systems are prepared for this type of usage and whether such queries can be processed efficiently by them. In our paper, we therefore investigate novel use cases for blockchain systems and elicit their requirements towards a data store in terms of query capabilities. We reflect the state of the art in terms of query support in blockchain systems and assess whether it is capable of meeting the requirements of such more sophisticated use cases. As a result, we identify future research challenges with regard to query processing in blockchain systems.

Keywords: blockchain systems; query processing; data models; data structures; block structures

Citation: Przytarski, D.; Stach, C.; Gritti, C.; Mitschang, B. Query Processing in Blockchain Systems: Current State and Future Challenges. *Future Internet* **2022**, *14*, 1. http://doi.org/10.3390/fi14010001

Academic Editor: Luis Javier Garcia Villalba

Received: 17 November 2021
Accepted: 13 December 2021
Published: 21 December 2021

Publisher's Note: MDPI stays neutral with regard to jurisdictional claims in published maps and institutional affiliations.

1. Introduction

Digitization fostered by the evolution of the Internet of Things (IoT) has made data one of the most important commodity in both business and private environments [1]. Data became the backbone for a variety of new data-driven application areas such as digital health [2], food supply chain [3], or the production of goods in Industry 4.0 [4]. All of these use cases have in common that they are permanently dependent on demand-driven data provisioning—i.e., the data generated and provided by several data producers must be made available to all data consumers in the required quality and quantity [5]. For this purpose, database systems are often used, as they significantly facilitate the management and provision of data [6].

However, due to the fact that data are nowadays highly valuable, they became attractive targets for cybercriminals who exploit these data in order to harm the involved parties. There is a wide variety of attack types, e.g., tampering with the data in these databases [7]. Detecting data tampering is nearly impossible without additional security measures, consequently being one of the most serious attacks to defend. Considering the worst-case scenario, where data are minimally tampered with, at stages that hardly arouse suspicion, can cause long-lasting damage to an organization [8]. These attacks are usually performed either by outsiders such as hackers, who are unaffiliated with the organization itself, or by malicious insiders such as untrustworthy database or system administrators. In traditional database systems, data are unprotected against this attack vector because they lack the necessary data integrity checks in the sense of ensuring that stored data are still in the same state as it was once inserted. Therefore, the recent emphasis lies on hardening these databases against data tampering [9].

Another problem is that attackers do not even have to attack the data itself to harm the involved parties. It is already enough to attempt to affect the availability of the data [10]. Denial-of-service attacks cause the database or the server on which it is running to become unreachable by flooding it with fake requests. While the server is occupied with processing the malicious requests, there are no more resources available for processing the legitimate requests, which is why they do not receive any feedback and thus, the data are no longer available to them. Another focus with regard to cybersecurity is, therefore, on strengthening availability to be prepared for the failure of resources [11].

Blockchains offer a solution to these two problems. Firstly, it is immutable and tamper-resistant, thus protected against data tampering. Secondly, it is decentralized and thus protected against denial-of-service attacks [12]. Yet, when Satoshi Nakamoto envisioned the first blockchain system for his digital currency *Bitcoin* [13], his priority was to solve the double spending problem, since there is no actual physical relinquishment in a digital currency. Therefore, many of the conveniences of traditional data management systems, such as a powerful query engine, are missing, i.e., they are much less convenient to use in terms of query language and query processing [14].

In this context, blockchains in particular offer many interesting additional use cases for queries due to their internal data management. In a blockchain, data are only appended, which results in the construction of a data log (i.e., blockchain data history) where different revisions of data coexist. This enables the possibility to query the data history for provenance analyses, unlike with a traditional database where data are modified in-place, which means that there is no natively existing data log to query [15]. The existence of this blockchain data history, however, means that applications are forced to store data externally to a blockchain and in many cases also need to perform additional query processing mostly local to the application.

This is why we investigate the necessary query capabilities for blockchain data histories. To this end, we provide three contributions in this paper:

1. Based on use cases from different application domains, we derive common types of usage of blockchain technologies in terms of types of data and queries.
2. For these types of data and queries, we investigate how they can be implemented in blockchain systems and how they can be supported by the available data history.
3. We explore the state of the art regarding query processing in blockchains and identify future research challenges.

By means of these three contributions, we identify open research gaps that need to be solved in order to enable efficient query processing in blockchain systems.

The remainder of this paper is structured as follows: We open by outlining the fundamentals of blockchain technologies in Section 2, with respect to their relevance in the context of this paper. In particular, our goal is to highlight the conceptual and architectural differences between blockchains and traditional database systems that are responsible for the challenges regarding efficient query processing. We then identify five emerging application domains in Section 3 where blockchains are becoming prevalent for data management. Based on a literature review, we identify types of data and queries that are relevant in these application domains. In Section 4, we generalize these types of data into two object types that must be distinguished when querying blockchains. Then, in Section 5, we determine for these two object types which query capabilities are required in blockchains to be efficiently usable in the application domains. In Section 6, we present the state of the art in research and discuss to which extent it provides these required query capabilities. Subsequently, we identify future research challenges in Section 7 and conclude this paper in Section 8.

2. Fundamentals of Blockchain Technology

Before we can delve into queries to a blockchain system, we need to address a few fundamentals of blockchain technology that have an impact on query processing. Even though it is often referred to as "the blockchain", a blockchain is actually a modular assembly of different components. In general terms, a blockchain is a ledger of sequential blocks that contain arbitrary information. This ledger is managed by a network of computers. That is, the distinctive feature of the blockchain is not what can be done with it—i.e., the secure management of data—but rather how this can be accomplished in a decentralized manner on a trustless infrastructure. For this purpose, well-established technologies from different fields of information technology are used in a blockchain. A blockchain architecture therefore has a modular structure, consisting of at least three layers: ❶ Storage, ❷ Network, and ❸ Consensus. Each layer is freely configurable to the respective requirements from a variety of technology variants, with all their advantages and disadvantages [16]. Figure 1 shows this modular architecture. In the following, we discuss these layers in detail.

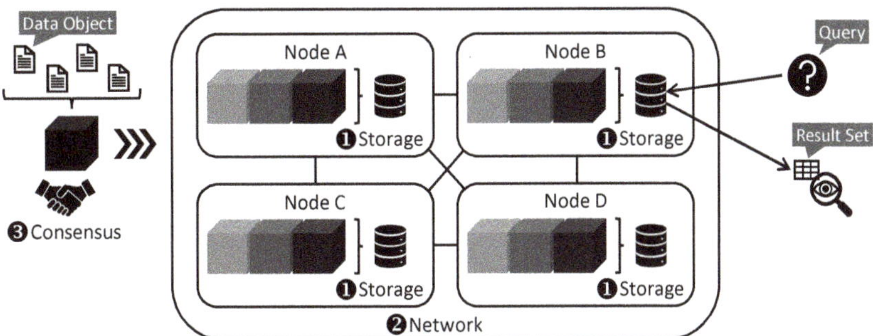

Figure 1. Simplified architecture of a blockchain system with its three layers: ❶ Storage, ❷ Network, and ❸ Consensus.

A blockchain is a list of blocks that are singly linked backwards using cryptographic signatures, with each block containing data. Backward linking is accomplished by including a header in a block that contains the hash value of its predecessor in addition to the actual payload data. A block cannot be modified subsequently, i.e., it is immutable. In particular, data and even entire blocks cannot be deleted retroactively due to this structure. In other words, a blockchain is an append-only data store. When new data are to be added to the blockchain, a new block must be created for this purpose, which is then appended to an existing blockchain [17].

There are many ways to manage a blockchain ❶. Usually, the data in a block are stored in a data structure that enables efficient verification of its integrity (e.g., *Merkle trees* [18], *Modified Merkle Patricia trees* [19]), and the blocks themselves are stored as a log-like structure on a storage device, with derived information stored in a state database for ease of access. The log is therefore mainly used to rebuild or verify the state database in case of problems [20].

Since data are never deleted from a blockchain, a blockchain automatically maintains a native data history. In contrast, a traditional database system must either manually implement the data history at the application layer (e.g., by implementing triggers to populate an audit trail table) or utilize specialized features like plugins for data history support (e.g., *Oracle Flashback Technology* (see https://www.oracle.com/database/technologies/high-availability/flashback.html; accessed on 15 December 2021) for *Oracle Databases* (see https://www.oracle.com/database/; accessed on 15 December 2021)) [21].

A blockchain is represented by multiple blockchain instances hosted on separate nodes in a distributed manner ❷. This replication approach increases availability and reliability [22].

In order to add new data to a blockchain, a new block must be created and announced to all nodes to become part of all blockchain instances. This distribution feature, however, leads to the possible situation where there could be competing blocks that are linked to the same predecessor and therefore cannot both be appended to the blockchain. To solve this, all nodes have agreed on a consensus mechanism ❸. This ensures that the network agrees on the next state of the blockchain, i.e., which block will be appended next to the blockchain. The consensus mechanism also defines, if a blockchain is permissionless or public, i.e., everyone can maintain a node, or if a blockchain is permissioned or private, i.e., only invited entities can maintain a node [23].

Permissionless. Consensus is typically achieved through communication (e.g., voting quorums). In a permissionless blockchain, however, the participants are unknown, so it is not even known how many are participating at all. Here, communication is replaced by computation. It requires that enough work has been put into the creation of a new block so that it can be appended to a permissionless blockchain, e.g., *Proof-of-Work* [24]. This ensures that only one participant generates a new block in a given period of time on average.

Permissioned. In a permissioned blockchain, the participants are known, and their number may be limited so that consensus can be reached through communication. This type of consensus is more lightweight and efficient. In most cases, participants do not trust each other, so a central database system as an alternative solution is not an option.

In summary, a blockchain has the following three key properties:

I.   It is **immutable**: Once a block is created, it is final. It cannot be modified subsequently, not even the link to its predecessor. The blockchain is an append-only data store. A new block can only be appended to an existing blockchain.
II. It is **tamper-resistant**: The data of a block are stored in authenticated data structures. These data structures are capable of verifying the integrity of their content. Tampering with their content gets therefore detected.
III. It is **decentralized**: Each node in a blockchain network manages its own instance of the blockchain. Thus, there is no single point of failure or attack. A consensus mechanism ensures that all nodes append the same, new block to the blockchain.

Although blockchains are becoming more and more popular as a secure and trusted data store, they differ significantly from traditional databases because of their completely different focus. While traditional databases are based on client-server architectures, blockchains are managed by a network of peer nodes, each of which holds a redundant copy of the full blockchain data. By eliminating the central management entity that has full control over the data store (and thus the data), trust is built—even if there is no trust among the participants of the network—but the management and communication over-head increases significantly. Besides this transparency, blockchains also create additional trust due to the immutability of the data and their tamper-resistance. These two properties are inherently guaranteed by the design of the blockchain, i.e., by organizing the data into blocks, all of which are linked via the cryptographic hashes in their headers. These blocks have no semantic meaning—they only reflect the chronological sequence in which the data are inserted into the blockchain. Data within a block can be entirely heterogeneous. There is no partitioning of the data into semantically associated tables or a strict schema for describing the data, as is the case for traditional databases. Meanwhile, traditional databases do not have comparable inherent security mechanisms. Yet, these security features are obtained in blockchains by the fact that they are append-only data stores, i.e., data cannot be subsequently deleted or modified. An update to an existing data record must be realized as a new entry, e.g., as a newer version of the complete data record or as an addition entry containing only the changes. As a result, blockchains cannot provide full CRUD support (create, read, update, and delete). However, due to the append-only structure of blockchains, they provide a complete data history in addition to the current state of the stored data, whereas traditional databases usually only contain the latest snapshot of the data. Depending on the chosen consensus mechanism, it may take some time until a data record is actually included in the blockchain, whereas there is no such delay in traditional database systems [25,26]. For all these conceptual and architectural reasons, query performance is also much higher for traditional databases in terms of throughput, the key efficiency metric for data stores [27,28]. Table 1 summarizes the main differences between traditional databases and blockchains that have an impact on their query capabilities.

Table 1. Main differences between traditional databases and blockchains.

Property	Traditional Database	Blockchain
Architecture	The traditional database model assumes that there is a central trustable administrator for the entire database. On that account, the database is hosted on a server and subordinated clients have to send their queries to this server.	The blockchain model assumes a network of equal nodes. Each node hosts its own instance of the entire blockchain. Although each node can execute queries independently, the network must agree on which is the valid state of the blockchain.
Replication	Even though traditional databases can use replication techniques internally, e.g., to prevent failure of physical storage media, externally there is only one database instance.	In a blockchain, there is full replication of all data on all nodes, i.e., the failure of a single node does not affect the availability of the data.
Validation	Traditional databases only ensure that if the database was in a consistent state before a write operation, it is also consistent after that operation. In addition, it is ensured that no side effects can occur when several users operate on the database.	Two types of validation take place in blockchains: (a) The nodes in the network agree in a consensus feature on what the valid state of the blockchain is, i.e., what data are part of the blockchain. (b) Users can verify the integrity of the data due to the tamper-resistance.
Structuring	Traditional databases organize data into tables, each with its own schema.	Blockchains organize data into blocks that have no semantic meaning.
Operations	Traditional databases provide full CRUD support.	Blockchains support only read and write (add new data) operations.
History	Traditional databases contain the latest snapshot of the data only.	Blockchains provide the complete data history.
Insertion	Inserted data are immediately available in a traditional database.	Due to the consensus mechanism, data are inserted with a time delay.
Performance	Traditional databases are geared towards a high data throughput.	The data throughput is significantly low due to the consensus.

Unlike in traditional database systems, data do not necessarily have to be stored in a blockchain. To this end, there are basically two approaches [29]. In the first approach called "on-chain", the actual data are stored within a blockchain. In the second approach, called "off-chain", the actual data are still stored in a traditional database system, but the information required to verify the actual data is stored on the blockchain. However, the verification overhead is significantly greater than with the first approach. Hybrid approaches are also possible, e.g., data are stored partly in a blockchain and partly in a traditional database system with their verification information on a blockchain.

Overall, the public verification of the data in a blockchain is a fundamental characteristic of blockchain technology. This transparency enables every node to check the integrity of the data in a blockchain, thus creating trust in the stored data. The focus on blockchain technology is on security, unlike traditional database systems, which focus on performance (i.e., transaction throughput). Additionally, blockchain technology provides protection against attackers, whether from hackers or malicious insiders, as well as protection against a single point of failure or attack, as data are replicated by many nodes, hopefully located around the world.

3. Application Domains Identified through Literature Review

As shown in the previous section, blockchains are technically very different from traditional databases, yet blockchains can in principle be used just like traditional databases—as a data store. Due to their decentralization, immutability, and tamper-resistance, blockchains offer additional security features that traditional data stores lack. At the time of this writing, many entities like companies, governments, and startups are evaluating the applicability of blockchain technology in their domains. As a result, further use cases utilizing blockchain technology in addition to a cryptocurrency have emerged over the course of time. According to Lo et al. [30], the use of blockchain technology is particularly beneficial when one or more of the following requirements are present:

- There is a need for establishing a trustworthy foundation between several parties without having to involve external authorities (e.g., notaries).
- There is a need for a single view of the truth (e.g., when different companies have to share data).
- There is a need for greater auditability by stakeholders through transparency (i.e., all published data are visible to every participant in the blockchain network) and provenance (i.e., the full history of data is available).
- There is a need for data being immutable (i.e., already stored data cannot be subsequently modified or deleted) and tamper-resistant (i.e., preventing an attacker from manipulating stored data).

From our literature review, we have identified five main application domains where one or more of the aforementioned requirements are present, and blockchain technology could therefore be a suitable technical design choice. These domains are *health data management* (see Section 3.1), *financial accounting* (see Section 3.2), *registries* (see Section 3.3), *food supply chains* (see Section 3.4), and *e-voting* (see Section 3.5). From these application domains, we derive typical types of data and types of queries in order to determine whether today's blockchain technology provides comprehensive query capabilities of the data history of a blockchain. These application domains are just a few selected examples that seemed particularly relevant in the context of our work. There are many other application domains that have similar query requirements, e.g., in the domains of *Smart Grids* [31,32], *digital rights management* [33,34], or *Smart Traffic* [35,36].

The main findings regarding the requirements for the query engine resulting from these use cases are summarized at the end of this section (see Section 3.6).

3.1. Health Data Management

In the health sector, digitization of many processes can significantly facilitate the lives of patients and physicians [37]. To this end, data in the form of patient records, e.g., electronic health records [38], must be shared and extended reliably and trustworthy among physicians. For example, a primary care physician prepares a medical record, and then refers the patient to a specialist, who adds their diagnosis. In addition, due to the *Quantified Self Movement* [39], people started to monitor themselves using IoT technologies, e.g., blood glucose measurements via continuous glucose monitoring [40] or heart rates via a smartwatch [41]. All these measured data are gathered in a central hub (e.g., a smartphone) and linked to compose a personal health profile [42]. By adding these personal health profiles to the patient records, physicians have access to even more health-related data which helps them to make a more accurate diagnosis.

The use of blockchain technology is suitable in such a use case because it allows decentralized data sharing. With a blockchain, a hospital can provide a data infrastructure through which physicians can share patient data with each other in a simple manner [43]. Moreover, the inherent immutability and tamper-resistance characteristics of a blockchain ensure data security, which is mandatory for medical data. This is particularly important due to the increasing threat of cyberattacks in healthcare [44]. By enabling patients to participate in the blockchain, they are empowered to provide additional health-related data on their own [45].

Especially when sensitive data such as health data are stored in a blockchain, it is obvious that data privacy protection measures have to be applied. This, however, contradicts the fundamental principles of a blockchain, according to which every participant has full access to all data. To this end, Peng et al. [46] present an approach in which data are stored tamper-resistant in a blockchain, but in which queries are processed in a privacy-preserving manner, and in which the result sets do not allow further inference about the data subjects.

There are multiple examples in literature in which blockchains are used to manage and share health data, e.g., De Aguiar et al. [47], Hasselgren et al. [48], Khatoon [49], Przytarski et al. [50], and Tanwar et al. [51].

Based on this research, we can conclude that there are two types of data in health data management:

- The data entered by physicians are usually documents, e.g., diagnosis and treatment plans, that are modified over time.
- The data entered by patients are usually measurements carried out by medical IoT devices that are only valid at a specific point in time.

In the context of health data management, queries regarding the current health status of an individual patient, information on disease progression over a given period of time, as well as aggregate measurement data are particularly relevant. Typical queries therefore include, but are not limited to:

- Retrieve all diagnoses of a specific patient from a given date.
- Retrieve the latest diagnosis of a specific patient where changes to the document are highlighted.
- Aggregate the measurements of a specific patient over a given period.

3.2. Financial Accounting

Today's accounting is still based on the double-entry system that was described in a treatise written by Luca Pacioli over 500 years ago [52]. The double-entry system has two sides known as debit and credit. Each financial record is entered into an account on both sides where the entry on the credit side is a corresponding and opposite entry of the debit side. The books are considered trustworthy if and only if the sum of the debits equals the sum of the credits [53]. Since a company is accountable to multiple parties—e.g., owners and investors—it is necessary to publish financial statements regularly. This implies that financial data must be shared with these shareholders, but also with tax advisors and financial authorities. The exchange of data is usually carried out via the error-prone import and export functionality of accounting software. As financial records must be immutable by law—i.e., they must not be tampered with retrospectively—such a modus operandi entails a considerable threat potential [54].

Since blockchain technology has already proven to be a backbone for cryptocurrencies, they also seem suitable for financial accounting. Accounts for any kind of assets, liabilities, equity, revenue, and expenses are established [55]. As all transactions between these accounts are transparent to all participants of the blockchain and no party has sole control over the blockchain due to its decentralized and distributed design, it can be considered a trusted single view of truth. Moreover, due to the immutability of financial records, a blockchain-based financial accounting is almost immune to tampering [56].

There are multiple examples in literature in which blockchains are used to support accounting, e.g., Faccia et al. [57], Gökten and Özdoğan [58], Schmitz and Leoni [59], Sveistrup Søgaard [60], and Zhang et al. [61].

Based on this research, we can conclude that there is only one type of data in financial accounting:

- The data entered by companies and tax advisors are financial records that are only valid at a specific point in time.

In the context of financial accounting, queries regarding the aggregated characteristic values over a given period of time or queries that support an accounting report are particularly relevant. Typical queries therefore include, but are not limited to:

- List all financial records for a given period (e.g., usually for a day, week, month, quarter, or year).
- Generate an accounting report by aggregating the financial records grouped by accounts for a given period.

3.3. Registries

A registry is an authoritative data source of records, usually maintained by a government agency. For instance, a land registry specifies who is permitted to use land, for how long, and on which conditions. Although the registry is maintained by a central authority, several other parties have to have access to the data in order to enable an economic and healthy business environment for the sale and purchase of property [62]. Only a few countries maintain a functioning land registry, which is still often based on paper-based documents, leaving them vulnerable to loss, misuse, or corruption. As a result, delays in ownership transfer or tampering with the land register are possible and bound to happen on a regular basis [63]. Another problem is that some registries exist duplicated in siloed entities so that this fragmentation might cause possible data conflicts and therefore, no single view of truth [64].

It is obvious that the use of blockchain technology can also provide a solution to all of these problems. On the one hand, blockchain technology ensures that documents are available to all participants almost immediately after they have been added to the blockchain. This eliminates unnecessary delays in processing that occur when paper-based documents are shipped. As a result, all participants always have the latest state of a document at their disposal and conflicting copies of one and the same document cannot exist [65]. On the other hand, the use of blockchains reliably prevents the forgery of documents due to the characteristics of a blockchain, i.e., its immutability and tamper-resistance. Since no central authority can gain full control over the blockchain, corruption is also not a problem as long as the majority of the participants are honest [66]. Obviously, it must be ensured that insights from the documents are not made public. However, this can be achieved by means of access policies and tailored permissions restricting the access of individual parties to the data. Such an approach is acceptable in terms of fraud protection as long as the blockchain itself is still governed by multiple entities [67].

This benefit is also demonstrated by many research papers for other registries, e.g., Benarous et al. [68], Rosado et al. [69], Sahai and Pandey [70], Shinde et al. [71], and Singh Yadav and Singh Kushwaha [72].

Based on this research, we can conclude that there is only one type of data in registries:

- The data entered into registries are usually documents (i.e., semi-structured data) that are modified over time. Typically, the latest state of a document is of importance, but in cases of conflicts, its history is also required (e.g., in court).

In the context of registries, queries regarding the latest of a certain document (as well as its history) are particularly relevant. Moreover, a data subject can be part of multiple registries, e.g., one registry containing all house owners and one containing all vehicle owners. In order to determine all properties of a certain data subject, a join between all available registries is required. Typical queries therefore include, but are not limited to:

- Retrieve the latest state of a specific document.
- Retrieve the latest state and a prior state of a specific document to highlight changes in the latest state.
- Join two or more registries on a certain attribute to get a holistic view of all stored documents.

3.4. Food Supply Chains

A supply chain is a network of entities in such sectors as agriculture and manufacturing ranging from producers, who produce a product or service, to the final consumer. In such supply chains, not only the physical exchange of the goods is important, but also the exchange of information about these goods. This information must be available to the participants of supply chain management in order to be able to ensure comprehensive quality control [73]. In the food industry, for example, meat products must maintain a cold chain in order to avoid endangering consumers' health [74]. This means, the temperature of the meat products has to be permanently monitored and fully documented during transport from the slaughterhouse to a retail store [75]. In order to exclude human errors, IoT technologies can be used for the metering and documentation [76].

While the use of IoT technologies can prevent unintentional measurement errors, it is also necessary to prevent tampering with the documents retrospectively, e.g., to guarantee that a breach of the cold chain is recognizable. Although the captured data must not be edited subsequently, it has to be possible to modify the accompanying documents to the meat products nevertheless, e.g., if additional entries are made during customs inspections or when the goods are handed over to the next supply chain entity [77]. The use of a blockchain to establish an immutable and decentralized data store for this data therefore makes sense. Besides eliminating the risk of fraud, the transparent data sharing capabilities of the blockchain also increase consumer confidence in the quality assurance of food products, as they can verify it in a tamper-proof manner [78].

There are multiple examples in literature in which blockchains are used to store proofs and certificates regarding food supply chains, e.g., Duan et al. [79], Köhler and Pizzol [80], Kayikci et al. [81], Shahid et al. [82], and Zhang et al. [83].

Based on this research, we can conclude that there are two types of data in food supply chains:

- The data generated by IoT devices are events and thus, only valid at a specific point in time (e.g., temperature or location).
- There may exist accompanying documents (i.e., semi-structured data) to the goods that are modified over time (e.g., during customs inspections).

In the context of food supply chains, queries that provide an aggregated overview of all captured data, as well as comprehensive querying of all documented data related to the transport, are particularly relevant. Typical queries, therefore, include, but are not limited to:

- Aggregate the events by specific attributes for a given period.
- Retrieve the latest state of an accompanying document for a given transport.
- Retrieve the latest state and a prior state of a specific document to highlight changes in the latest state.

3.5. E-Voting

Electronic voting systems (known as e-voting) are a means of strengthening democratic processes. By digitizing the election process, not only is bureaucracy reduced, but people can cast their votes much more efficiently. This is an advantage especially for elderly voters or voters with a disability, as e-voting enables them to participate in the election without having to leave home and rely on the help of others [84]. While in the past, mostly technical difficulties impeded the introduction of e-voting, in today's fully connected world, it is rather a matter of security concerns [85]. To this end, the transmission of votes must be trustworthy and secure [86], and the secrecy of the ballot has to be respected [87].

However, one of the most important confidence-building measures is to ensure full transparency in e-voting and election results. This means, all voters must be able to verify that every vote is counted and that ballots are not manipulated retroactively. The use of blockchains is therefore particularly suitable to manage the votes. First of all, the community decides by consensus which data are included in the blockchain, i.e., which votes are valid. Storing votes in a blockchain ensures that they are immutable, and tampering can be detected immediately. In addition, blockchains provide great transparency because each participant in the blockchain network keeps a complete copy of the blockchain—and thus all of the data—on their node [88]. Furthermore, the decentralized nature of blockchains ensures availability, as they are less susceptible to denial-of-service attacks than a centralized approach [89].

There are multiple examples in literature in which blockchains are used to support secure and transparent e-voting, e.g., Hanifatunnisa and Rahardjo [90], Hjálmarsson et al. [91], Kshetri and Voas [92], Ruparel et al. [93], and Wang et al. [94].

Based on this research, we can conclude that there is only one type of data in e-voting:

- The votes are stored in the blockchain as independent records. Once a vote has been cast, it must not be subsequently altered or deleted. Without any loss of generality, we assume that some kind of verification of whether a ballot is valid takes place before the votes are entered into the blockchain. Therefore, no extensions to the stored data are required.

In the context of e-voting, statistical queries that aggregate the stored data are particularly relevant. Typical queries therefore include, but are not limited to:

- Determine the final result of an election.
- Determine the voting behavior of different groups of voters.
- Determine which shifts of voters happened compared to the last election.

3.6. Lessons Learned

Derived from the presented application domains, we conclude that there are two different types of data that are entered into a blockchain. We outline their characteristics in Table 2. The first type of data entered into a blockchain is only valid at a specific point in time, which we call a *constant object*. Constant objects are, in other words, just events, such as those known from complex event processing [95]. However, there is a peculiarity in dealing with the timestamp of a constant object. This is because the timestamp can be dependent on the block in which the object is stored (i.e., an object with a *block-dependent timestamp*), or dependent on the object, because the object itself provides a timestamp attribute that must be used rather than the timestamp of the block (i.e., an object with an *object-dependent timestamp*). The second type of data entered into a blockchain is modified over time, which we call an *expandable object*. As the modifications are scattered over many blocks, they must first be combined in order to be used further. Therefore, expandable objects have only block-dependent timestamps. We use the term "object" to describe a set of attributes, i.e., data in the form of a set of key-value pairs, so-called fields. Although the concept of objects is mainly known in the paradigm of object orientation, this data model does not restrict us to the use of object-oriented data stores. These objects can also be represented in other data models such as *JSON documents*, *RDF triples* (i.e., mapping the fields of an object to individual triples), or *XML instances*. Listing ?? shows an object named `obj1` with three attributes and their values in those three representations. We discuss those object types further in Section 4.

Furthermore, from the presented use cases, we derive eight query capabilities that an efficient query engine for blockchain systems has to support in order to be usable in the given application domains. These required capabilities are *projection*, *selection*, *sorting*, *aggregation*, *grouping*, and *joins*. These operators are well-known from the *relational algebra*, on which the query languages of many traditional database systems are based.

Table 2. The two types of objects which are relevant in the context of blockchains.

Type of Data	Main Property	Timestamp
Constant Object	This type of data is only valid at a specific point in time. Thus, it is final, i.e., its fields (key-value pairs) and corresponding values are constant and do not change over time.	block-dependent object-dependent
Expandable Object	This type of data can grow and shrink over time, i.e., in the future, new fields may be added, values of existing fields may be modified, or existing fields may be removed. Any state of the object can be restored by exploiting the history feature of the blockchain.	block-dependent

Listing 1. An object with three attributes and their values represented as a JSON document, RDF triples, and an XML instance.

An object represented in the form of a JSON document named `obj1`.

An object represented in the form of RDF triples.

An object represented in the form of an XML instance.

```
{
"attr1": "val1",
"attr2": "val2",
"attr3": "val3"
}
```

```
<obj1, attr1, "val1">
<obj1, attr2, "val2">
<obj1, attr3, "val3">
```

```
<obj1>
<attr1>val1</attr1>
<attr2>val2</attr2>
<attr3>val3</attr3>
</obj1>
```

Projection means selecting specific attributes from objects that are included in the result set, i.e., if an object has several attributes, only a specific subset of them is returned. For instance, a physician requires a projection operator to query specifically blood pressure measurements from an electronic health record, which also includes other medical data such as blood glucose measurements or dietary studies. Selection means eliminating objects from the result set, i.e., an object is only included in the result set, if its attribute values meet a given condition. For instance, a physician requires a selection operator to query for female patients (i.e., patients whose attribute "gender" is set to "female"). Sorting means to sort the objects in the result set in ascending or descending order, based on the values of the attributes of the objects. For instance, in financial accounting, it is necessary to sort the accounting items in order to present them according to the date they were registered. Aggregation means to compute a single value from a set of values with the help of an aggregate function, such as average, maximum/minimum, or sum. For instance, in financial accounting, an aggregation is required to compare the total sum of income with the total sum of expenses in the end. Grouping means to partition objects into groups of objects, based on the values of their attribute. For instance, land registries have to group the landowners based on the county their property is assigned to. Usually, an aggregation is then applied on these groups, e.g., to determine how much real estate tax each county receives. Joining means to combine data from multiple sources into a joint result set. While in traditional database systems joins are applied to different tables within the same database, in blockchains there is no such semantically structuring construct like a table. Therefore, joins have to be applied to different blockchains. This, however, raises further technical issues, see Sections 5 and 7. Nevertheless, there are use cases in which joins have to be supported by blockchain systems. For instance, if there are different registries, e.g., a land register and a car register, each stored in its own blockchain. In order to query all possessions of a data subject, a join on all of these blockchains is necessary.

In addition to these six basic query operators, which are also known from traditional database systems, blockchains have special requirements towards query capabilities due to the two different object types that have to be handled by them. Firstly, there are *temporal queries* when dealing with constant objects. In temporal queries, the temporal relationship of the data plays a key role. These time references can be obtained from two different sources: On the one hand, each block has its own inherent timestamp. Since new blocks can only be added at the end of the blockchain, the sequence of the blocks implicitly reflects the chronological order in which they were created. This timestamp is used for block-dependent objects for temporal queries. However, it is possible that this timestamp deviates substantially from the time at which a data object was captured, since data initially remain in a data pool until a consensus is reached, and they are added to a new block. Therefore, for object-dependent objects, where the time of capturing the data is crucial, an additional individual timestamp for each object is needed. For instance, in the e-voting context it is necessary to query only valid votes, i.e., only ballots that were submitted neither too early nor too late have to be considered. Secondly, there are state-based queries when dealing with expandable objects. Such objects are initially added to the blockchain and then changes are made by means of transactions (e.g., to change certain attribute values, and add or remove some attributes) which are also stored in the blockchain. In a state-based query, the complete change history up to a specific point in time must therefore first be retrieved from the blockchain in order to assemble the expandable object. For instance, in the food supply chain it must be possible to query the status of a food product at any time between production and sale, e.g., in order to monitor the cold chain.

Table 3 provides an overview of these six basic operators as well as the two blockchain-specific query capabilities. More details on these query options are provided in Section 5.

Table 3. Overview of the six basic query operators (white rows) and two blockchain-specific query capabilities (gray rows) derived from the presented application domains.

Query Capability	Main Property
Projection	It is possible to specify which fields (i.e., key-value pairs) are included in the result set.
Selection	It is possible to specify which objects are included in the result set.
Sorting	It is possible to sort the result set by given fields.
Aggregation	It is possible to aggregate the values of certain fields using functions.
Grouping	It is possible to group given fields.
Join	It is possible to join different blockchains.
Temporal Queries	It is possible to query constant objects based on a timestamp. While for block-dependent objects there is an inherent timestamp given by the block they are stored in, object-dependent objects have their individual timestamp, which is specified in their attributes.
State-based Queries	It is possible to query expandable objects. Expandable objects can be scattered over multiple blocks, meaning that a state-based query must first find all pieces and compose them.

4. Object Types in Blockchains

From the presented application domains in Section 3, we derive two object types that are relevant in the context of blockchains, namely constant objects and expandable objects. Their main properties are summarized in Table 2. In the following, we elaborate on these two object types and describe why they need to be considered in particular when managing data in blockchains.

As described in Section 2, blockchains are append-only data stores where blocks are appended to an existing blockchain. Furthermore, blocks cannot be modified subsequently, so the data within a block are immutable. If changes to the data must occur, there are two options. Either the complete object with all its fields is recreated or only a change history is kept. This means that there are two different forms of use. Either, an object *lives* until a new version of it is added to the blockchain or the entire change history of an object must be searched in the blockchain and applied to the *genesis object*, i.e., the original version of the object. These two forms of use are reflected by the following two object types:

Constant Object. A constant object is final. This means that once the object is added to a block, its fields do not change. Constant objects occur over time and are valid at a specific point in time. In other words, constant objects are events, i.e., actions or occurrences that happened at a specific point in time.

Expandable Object. An expandable object is never final. This means that over time, the fields of this object are modified, new fields are added, or existing fields are removed. In other words, expandable objects are documents that get modified over time.

Constant objects are, for example, votes in e-voting (see Figure 2a) or blood pressure measurements from medical IoT devices in health data management (see Figure 2b).

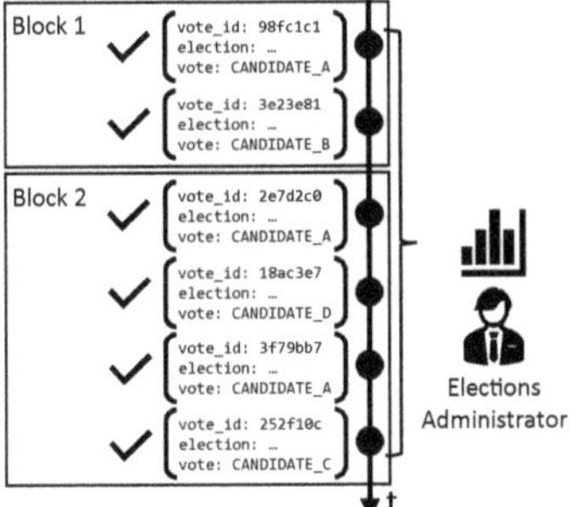

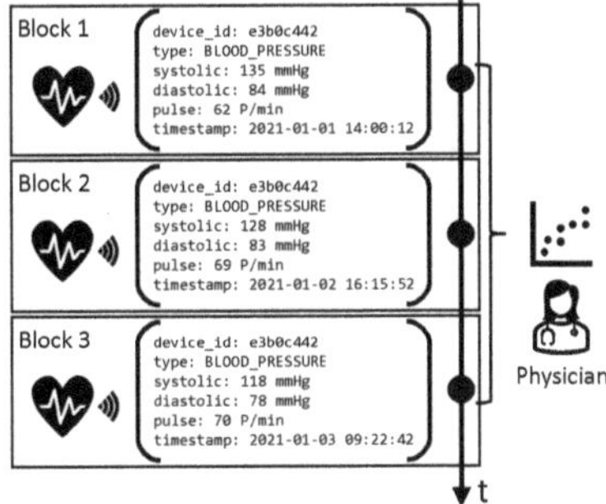

(**a**) Votes during an election, represented as constant objects with block-dependent timestamps, which are aggregated by an election administrator.

(**b**) Blood pressure measurements from medical IoT devices, represented as constant objects with object-dependent timestamps, which are analyzed by a physician.

Figure 2. Two different use cases utilizing constant objects with (**a**) block-dependent timestamps and (**b**) object-dependent timestamps.

In e-voting, votes are created by voters during elections. These votes are only valid once they are successfully added to the blockchain. A vote does not contain its own timestamp attribute, because in this case, only the timestamp of the block is relevant. An election official can query and aggregate these votes to derive valuable information about an election. For these queries, it is relevant in which block a vote is included.

In health data management, a medical IoT device performs blood pressure measurements at certain time intervals. These measurements are either added to the blockchain individually or in batches. A measurement contains, among other attributes, a timestamp that records the time of the measurement. A physician can query and aggregate these measurements so that valuable information can be derived for the patient. For these queries, however, it is not relevant in which block the measurement is included, but at which time it was performed (nota bene: Due to the delayed insertion of data into the blockchain, not only the timestamp of a measurement can significantly differ from the timestamp of the block it is stored in, but also the chronological order in which measurements are captured can differ from the order within the blocks.).

Thus, in the first example, the timestamp of the block is relevant, but in the second example, the timestamp of the object is relevant. For this reason, we introduce the following notion for timestamps on objects:

Block-Dependent. In this case, the object depends on the timestamp of the block it was included in. Each block has its own timestamp, i.e., the time at which it was created. Here, the timestamp of a block acts as a global timestamp for all its payload data, superseding possible timestamp attributes of objects, thus all objects in a block have the same timestamp.

Object-Dependent. In this case, the object has its own timestamp attribute. Additionally, it is not relevant in which block this object was included. During query processing, the timestamps of these objects must be considered instead of the timestamp of a block. However, this entails new challenges. In a blockchain architecture, there is no guarantee that the objects are sorted by the timestamp attribute of the objects. As a result, when searching for an object with a specific timestamp, it can only be assumed that the object was created earlier than the block that includes it. Thus, the lower search bound is set by the timestamp of the object, however, no statement can be made about the upper search bound.

Whether an object has a block-dependent or an object-dependent timestamp is determined by its further usage. In our e-voting example, the action is to cast a vote and this is considered to be performed once it is correctly added to the blockchain. In our health data management example, the action is a blood pressure measurement carried out by an IoT medical device, which is considered to be performed once the measurement is successfully completed. This action is completely independent of the creation of a block for a blockchain.

Expandable objects are, for example, documents in land registries (see Figure 3). An expandable object consists of a genesis object (i.e., the source object) as well as modifications to the object that are scattered over numerous blocks. As a result, it has as many states (i.e., document revisions) as how many blocks exist that include fields of this object.

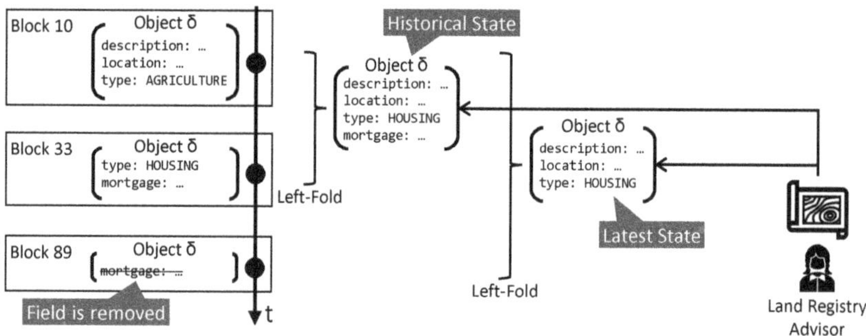

Figure 3. A land registry document, represented as expandable object, which is modified over time. Different states of the document can be retrieved, i.e., the latest state and all historical states.

In land registries, land documents are inserted, modified, and deleted over time. When a land document is modified, it means that fields of the document are modified, new fields are added, or existing fields are removed. The result of a modification is a new state of the expandable object. Thus, each block that include a modification of an expandable object represents a different state of this very object. A land registry advisor can query these land documents at any available state. For this, the requested state of the document has to be computed.

To compute a state of an expandable object, all fields from the previous and the requested block must be combined. This is done by recursively recombining the fields from the first block that includes fields of the object until the requested state—this approach is also called *left-folding*.

In our land registry example, the object first appeared in Block 10, the so-called genesis object. After that, there have been two modifications to it, namely in Block 33 and Block 89. This means that there are three states for this object, all of which can be queried. Querying its state in Block 10 is simple, since no modifications have taken place yet. Querying its state in Block 33 requires its assembly by combining the fields from Block 10 with the modifications in Block 33. The same procedure is used for querying its state in Block 89, although an additional combination step has to be performed then.

Furthermore, the timestamps of expandable objects are only block-dependent, i.e., the block defines the corresponding timestamp for these objects.

5. Query Capabilities for Blockchain Technology

As discussed in Section 4, the different object types have a significant impact on how data can be queried in a blockchain. Therefore, in this section, we adapt a potential query language to the object types using the query capabilities listed in Table 3 and elaborate on possible issues that need to be considered when implementing a query engine.

In blockchain technology, writing data is a completely different process compared to traditional database systems. This is due to the consensus mechanism used to add new data to the blockchain (see Section 2). Therefore, we only consider non-modifying query techniques, i.e., read queries as they have no persistent effects. Nevertheless, data can still be added to a blockchain by creating a new block that includes the new data and propagating it via the given consensus mechanism.

A query engine consists of a frontend and a backend. The frontend is responsible for transforming a query written in a defined query language into an intermediate representation. The backend is responsible for processing this intermediate representation and computing the result of that query.

The use cases shown in Section 3 require comprehensive query capabilities such as aggregations or joins. For the complete breakdown of required capabilities, see Table 3. We consider a query engine to be powerful, if it supports a query language with at least the same power as a SELECT statement from the declarative query language SQL—just like in traditional database systems. Current blockchain systems, however, have native but naive query interfaces [96]. Moreover, their query languages and the efficiency of query processing is severely limited [97]. Since descriptive query languages have proven themselves in practice also for object-oriented database systems [98], we describe the required queries in SQL. SQL provides an expressive query language [99], however, SQL is just one example that can easily be replaced by any other declarative query language. In particular, we focus on the SELECT statement, since this is used for the read queries. However, the SELECT statement cannot be simply adopted, but has to be modified to support the different object types.

In relational database systems, the SQL SELECT statement is the most common option to query a database. Within this SELECT statement, there are various clauses intended for, e.g., selecting, aggregating, or sorting. Table 4 shows these various clauses and maps them to the respective query capability along with a mapping to the blockchain domain.

For almost all of these clauses, a relatively straightforward mapping to the blockchain domain can be found. However, the JOIN command represents an exception. Since blockchains have no logical internal structuring (nota bene: The blocks in which the data are organized have no semantic meaning regarding the data. They only represent the chronological order in which the data were added to the blockchain.) (e.g., in semantically and schematically homogeneous tables), a JOIN gets a different and new meaning in this context. As illustrated in the example of the registries (see Section 3.3), it happens in practice that data from a single data subject are contained in several different blockchains. To collect and combine all information, a JOIN across multiple blockchains is required. However, as outlined in Section 2, blockchains do not have a uniform structure. Thus, it must be resolved how a JOIN can be realized despite the highly diverse technologies that are involved in this case.

While these SQL clauses are sufficient to cover all six basic query operators (see Table 3), the inclusion of novel blockchain-specific object types (see Section 4) represent a significant deviation from SQL. Due to these object types, additional query capabilities—alongside with extensions to the query language—are needed in blockchain systems.

Table 4. The various clauses of an SQL statement and their mapping to the blockchain domain.

Query Capability	Relational Data Model	Blockchain Domain
Projection	`SELECT <columns>` An SQL statement starts with the projections, a list of columns to include in the final result set.	`SELECT <attributes>` Instead of columns, attributes of objects are specified.
Join	`FROM <table> JOIN <other tables>` This clause indicates the table from which to retrieve the data. `JOIN` subclauses enable the joining of additional tables.	`FROM <blockchain> JOIN <other blockchains>` Instead of a table, the blockchain is specified. If there is only one blockchain given, then the clause might be omitted. If there is more than one blockchain given, `JOIN` subclauses are required.
Selection	`WHERE <comparison predicates on columns>` This clause eliminates all rows from the result set where a comparison predicate does not evaluate to true.	`WHERE <comparison predicates on attributes>` Instead of rows, objects are eliminated.
Grouping & Aggregation	`GROUP BY <columns>` This clause groups values of one or more columns in conjunction with aggregation functions in the projection on those columns.	`GROUP BY <attributes>` Instead of columns, attributes of objects are specified.
	`HAVING <comparison predicates on groups>` This clause eliminates all groups of returned rows to only those whose comparison predicate does not evaluate to true.	`HAVING <comparison predicates on groups>` Instead of rows, objects are returned.
Sorting	`ORDER BY <columns>` This clause indicates the columns to use to sort the result set including the sort direction.	`ORDER BY <attributes>` Instead of columns, attributes of objects are specified.

Constant objects are self-contained, which means that, considered individually, they do not provide valuable information in most cases. Thus, it is suitable to consider several of these objects at the same time. This can be done, for example, either in the form of an aggregation or viewing the data as time series to track any trends. In order to support this, a start and end point are required. However, the range queries differ here in whether the objects have block-dependent or object-dependent timestamps. For objects with block-dependent timestamps, the timestamp of a block is relevant, therefore, it must be possible to specify two block numbers. Thus, it must be possible to search between block N_1 and block N_2. To apply this to SQL, the `SELECT` start clause could be adjusted as follows:

Block Range. `SELECT <attributes> BETWEEN BLOCK` N_1 `AND` N_2
   (where N_1 and N_2 of type `Integer` and $N_1 \leq N_2$)
   A *block range* is necessary when a blockchain stores constant objects with block-dependent timestamps.

The situation is different for objects with object-dependent timestamps. Here, the order in which the data was added to the blockchain is irrelevant, it only matters when the data was originally generated. Therefore, it is necessary to search via the timestamp of the objects. This means that only objects created between timestamp T_1 and T_2 are searched. To apply this to SQL, the `SELECT` start clause could be adjusted as follows:

Timestamp Range. `SELECT <attributes> BETWEEN TIMESTAMP` T_1 `AND` T_2
   (where T_1 and T_2 of type `DateTime` (e.g., ISO 8601 [100]) and $T_1 \leq T_2$)

A *timestamp range* is necessary when a blockchain stores constant objects with object-dependent timestamps.

Even though the two queries look very similar, they are internally very different from each other. Since the block range corresponds to the structure of the blockchain, such a query can be supported very efficiently. The timestamp range, however, requires all blocks between the block with timestamp T_1 (nota bene: Even if it is not clear when an object is added to the blockchain, the insertion time (i.e., the timestamp of a block) can in no case precede the timestamp of the object (i.e., the time of capturing).) and now to be searched, since the timestamp of the actual block where the object has been included is greater than the timestamp of the object itself.

Expandable objects have fields that are scattered over one or more blocks. These objects must be assembled before they can be processed to compute the result of a query. It is obvious that the states of all processed objects must be at the same *height* (In this context, the term "height" is used to describe the block number within a blockchain up to which all required objects have to be assembled.) to prevent the processing of incompatible states of data. Therefore, it is necessary to specify a block number N up to which block the objects are being assembled (nota bene: A lower bound is not required in this case, since it is always necessary to start with the genesis object and apply all modifications from there.). To support this, the SELECT start clause could be adjusted as follows:

Block Number. SELECT <attributes> ASOF BLOCK N
(where N of type Integer)
A *block number* is necessary when a blockchain stores expandable objects.

This way, all required query capabilities for all object types can be represented in a declarative query language. This shows how powerful a declarative query language is. However, the query language is just the frontend of a query engine.

The actual issues arise when the backend of a query engine is considered, as it accesses the underlying data structures to compute the result of a query. We identified the following eight issues that need to be addressed:

1. JOIN operators as provided by traditional database systems, do not need to be considered here, as there is no demand for this functionality in practice. Unlike traditional database systems, blockchain typically store data on a single topic only. An internal structuring into separate tables, each with its own schema, is therefore not necessary in blockchain systems. Consequently, joins cannot be performed within the data set of a single blockchain. However, there are use cases that require a join between data sets held in different blockchains. For instance, Blockchain X contains health data that are captured self-reliant by patients as part of the Quantified Self Movement, while Blockchain Y contains clinical data of these patients captured by hospital staff as part of health checks. In order to get a comprehensive view of a patient's health situation or history, physicians need to be able to join the data from these two blockchains. Since each blockchain system has its unique technical architecture regarding its storage, network, and consensus (see Section 2), such a join represents a substantial technical challenge.

2. Unlike a relational table, where all data are applied to a table schema, blockchain objects have no common well-defined schema. Here, the structuring of the objects is done solely at the application level. That is, each application stores its objects in its own predefined schema. However, when several applications share a blockchain to store their data, multiple schemas are simultaneously present in that blockchain. Therefore, the question is how this inhomogeneity affects query processing?

3. Data read from a blockchain should always be verified to detect any tampering. However, data could also be stored externally to a blockchain in a database system with better query capabilities, but without verification capabilities. Therefore, the question here is how, and when does the verification of the data take place? During query

4. Database systems utilize index structures to facilitate query processing. Can such structures also be used for query processing in blockchains? If so, how could these look like for constant objects and/or expandable objects? Is it possible to verify the data in these index structures?

5. Blockchains lack an internal structuring that has a semantic meaning. While the segmentation into blocks is beneficial for some queries—think of queries for expandable objects, for instance, where the state up to a specific point in time is required, which can be easily realized via a query on the block number—this complicates queries on the timestamp of a constant object, for instance, since all blocks created at this timestamp or later have to be traversed for this purpose.

6. The query processing of constant objects and expandable objects is very different. Can these objects be technically processed simultaneously in a blockchain? If so, does it make sense from a query language perspective?

7. The query processing of constant objects with object-dependent timestamps is more complex than that of constant objects with block-dependent timestamps. Can these objects be technically processed simultaneously in a blockchain? If so, does it make sense from a query language perspective?

8. The query processing of expandable objects is significantly more complex than that of constant objects with block-dependent timestamps, since for each object it is first necessary to determine which attributes it has, and in which blocks they are located.

6. Overview of the State of the Art

While blockchain technology was initially developed for cryptocurrencies, for which it is sufficient to query the current account balance, the new use cases identified from the application domains in Section 3 introduce different types of objects (see Section 4) that require comprehensive query capabilities (see Section 5). Since there is not a standard for blockchain systems, but rather a modular design that can be freely configured from a variety of technology variants (see Section 2), there are various blockchain systems, each targeting a different goal. As a result, the query capabilities of these systems are quite different. In this section, we therefore first consider the state of technology (see Section 6.1) and then the state of research (see Section 6.2) in the field of query processing in blockchains.

6.1. State of Technology

The currently most popular blockchain system *Hyperledger Fabric* [101] manages a ledger that consists of a blockchain and a database that holds the current *world state*. The world state represents the latest state of a blockchain and is stored in an additional NoSQL database. Hyperledger Fabric uses *CouchDB* (see https://couchdb.apache.org; accessed on 15 December 2021) to this end. Despite the fact that a blockchain maintains a native data history, however, there are only limited interfaces to access this data history (e.g., through Fabric SDK (see https://hyperledger-fabric.readthedocs.io/en/release-2.2/fabric-sdks.html; accessed on 15 December 2021) or smart contracts). It is possible to execute comprehensive queries against the CouchDB, which manages the latest state. However, the result of a query is not cross-checked against the blockchain, so there is a possibility of reading tampered data.

In such blockchain systems, there is no efficient technique to query the underlying data structure, i.e., the data history of the blockchain. A solution to overcome this limitation is therefore to duplicate the data of the blockchain (or even just the current state) into a separate database with support for a powerful query engine, while sacrificing the built-in technique of the blockchain to verify the integrity of data while computing the result of a query. If, thus, information must be directly extracted from the data history of the blockchain, expert knowledge and self-developed tools are required to extract this information.

Furthermore, there are systems that have features from blockchains and databases. They are called hybrid systems and there are two alternative approaches. The first approach is to start with a blockchain system and then enhance it with database features. The second approach is to start with a database system and then enhance it with blockchain features. Ruan et al. [102] compare six of such hybrid systems and came to the following findings:

Blockchain Systems Enhanced with Database Features. These systems use blockchains as an integrity-protected data store and utilize a separate database layer on top of it. Within the network, storage operations are replicated (e.g., a block containing transactions) rather than individual transactions. Examples are:

- *BlockchainDB* [103] provides a key-value database layer on top of a blockchain, which provides a simple get/put interface as well as an additional verify method for data verification.
- *FalconDB* [104] provides a "traditional" database layer with temporal attributes on top of a blockchain. It relies on smart contracts for querying as there is an incentive model that each node remains honest.
- *Veritas* [105] provides a verifiable database layer on top of a blockchain.

Database Systems Enhanced with Blockchain Features. These systems use ordinary database systems and utilize transaction-based replication. Within the network, each node manages its own database instance and executes globally ordered transactions (achieved through a consensus mechanism) on it. Examples are:

- *BigchainDB* [106] provides a blockchain layer on top of a *MongoDB* (see https://www. mongodb.com; accessed on 15 December 2021) database. As all blocks, transactions, and metadata are stored in it, the full query power of MongoDB can be used to query data.
- *Blockchain Relational Database* [107] integrates a blockchain layer into a relational database management system, namely *PostgreSQL* (see https://www.postgresql.org; accessed on 15 December 2021). PostgreSQL was chosen because it keeps all versions of a row. Usually, relational database systems update data in-place and maintain a rollback log.
- *ChainifyDB* [108] provides a blockchain layer on top of arbitrary database management systems that are SQL-99 [109] compliant. It uses a new processing model that reaches consensus on the effects, i.e., database states and snapshots.

We conclude that in both approaches, the system *can* provide query capabilities that are mostly as powerful as the query engines of the applied database systems (i.e., document-oriented databases and relational databases). However, these underlying traditional database systems provide no support for block range queries, timestamp range queries, and block number queries, as required in modern blockchain use cases. In addition, each approach has its own disadvantage.

The disadvantage of blockchain systems enhanced with database features is generally that data in the database are decoupled from the data in the blockchain so that verifying the results of a query is an additional step that can become expensive. Depending on how the data of the blockchain are stored in the database, queries are possible either only on the latest state or also on the history. FalconDB uses *MySQL* (see https://www.mysql.com ; accessed on 15 December 2021), which provides a relational data model, that they extended by temporal attributes to support SQL queries on the history.

The disadvantage of database systems enhanced with blockchain features is generally that the database system itself might not use tamper-resistant data structures so that tampered data is detectable. There are techniques to overcome this such as querying multiple nodes in the blockchain network and comparing the result or re-executing the transactions from the blockchain to detect incorrect data. However, these techniques are cumbersome and can also become expensive.

6.2. State of Research

Given the problems with the State of Technology, there is also a variety of research. These can be divided into four research directions:

Improvements to the Frontend Query Capabilities. *Ethereum* [110] is a popular public blockchain that supports smart contracts, which are programs with code (i.e., functions) and data (i.e., states) that run on the blockchain. It uses the key-value database LevelDB as persistent storage. Han et al. [111] extend the Ethereum-based blockchain system *quorum* by an embedded relational database system *SQLite* (see https://www.sqlite.org/index.html; accessed on 15 December 2021) next to *LevelDB* (see https://github.com/google/leveldb; accessed on 15 December 2021) enabling SQL SELECT queries. In this system, the data of smart contract transactions are stored in the SQLite database instead of the LevelDB database. Smart contract transactions can use SQL queries (e.g., range or conditional queries), which are then executed by the relational database system. However, there are some open questions, e.g.,

- What happens, if the data in the relational database SQLite is tampered with?
- Smart contracts in Ethereum only have access to the latest state of their data. Is this also the case here?

The research work of Tong et al. [112] also focuses on providing SQL support in blockchains systems. However, they take a different approach. They introduce an SQL middleware, which encapsulates RPC-based (remote procedure call) interfaces of blockchain systems as SQL interfaces to facilitate SQL queries on the blockchain data, just like the aforementioned approach, where blockchain systems are enhanced with database features. Furthermore, Li et al. [113] present a data query layer called *EtherQL*, which enables a set of useful analytical queries such as range and top-k queries on the blockchain Ethereum.

Efficiency Improvements in Query Processing. Bragagnolo et al. [114] use the parallelization technique Map/Reduce to extract and analyze information from a blockchain, in their case from the Ethereum blockchain. Here, a master node instructs different jobs to worker nodes, each of which extracts data from the Ethereum blockchain and writes them to a relational database. After that, queries can be made to the relational database to obtain information from the Ethereum blockchain.

Xu et al. [115] present an accumulator-based authenticated data structure that allows aggregation over arbitrary attributes. This enables lightweight users, i.e., users who have only the block headers locally stored, to have service providers storing the full blockchain to execute boolean range queries, while allowing them to verify the integrity of the results.

Xing et al. [116] present a subchain index structure for the transaction chain. Here, the transaction chain is divided into subchains and different subchains are linked with hash pointers. The goal is to shorten the query path for queries on historical transactions.

Jia et al. [117] present the *AB-M tree* structure as a storage structure for transactions, which combines the advantages of *balanced binary trees* (fast data retrieval) and Merkle trees (fast data verification). Instead of storing transactions in an ordinary Merkle tree within a block, they are now stored in an AB-M tree. This provides faster transaction retrieval, but at the same time guarantees the integrity of the transactions.

Peng et al. [118] and, based on this, Wu et al. [119] present a middleware layer called *Verifiable Query Layer* (*VQL*). It extracts information about the blocks, their transactions, and possible balances from an underlying blockchain system and stores these data reorganized in one or more databases so that queries can be answered more efficiently. Then, a cryptographic hash value for each generated database is computed and stored in a blockchain, preferably in the underlying blockchain system. Whenever data is queried through the middle layer, the integrity of the queried database can be verified by comparing the hash value in the blockchain with the hash value computed by the user.

Tailored Blockchain Optimizations for Specific Use Cases. As IoT technologies capture growing volumes of time series data, there is an emerging need to comprehensively analyze it in an efficient manner. While there are approaches to verify the authenticity of the sources of this IoT data [120] and subsequently provide these time series data to third parties on a demand-driven basis [121], it is also necessary to ensure that the data cannot be tampered with when it is stored and managed.

Wortner et al. [122] therefore investigate particularly for time series data how these can be managed in blockchain systems and how in particular their timestamps, which play a key role in subsequent analyses, can be protected against tampering. In this context, however, the focus is solely on the storage of the data. An efficient processing of queries or let alone an analysis of the blockchain data is completely out of scope. This is being researched by Dhanush et al. [123]. In their approach, however, the time series data must first be completely extracted from the blockchain and then stored and analyzed in a special time series database (e.g., *InfluxDB* (see https://www.influxdata.com/products/influxdb/; accessed on 15 December 2021)) for which there are tailored analysis tools and dashboards (e.g., *Grafana* (see https://grafana.com; accessed on 15 December 2021)). This causes a large overhead, because there are no efficient ways to restrict the amount of data in such a way that only those data are read that are relevant for the analysis. Since the amount of data in the blockchain is continuously growing due to the append-only nature of the blockchain, this overhead is also constantly increasing. Another problem with this approach is the fact that once the data has been extracted, there is no longer any protection against tampering. This completely undermines the main reason why the data was stored in the blockchain in the first place.

Yu et al. [124] therefore propose a novel blockchain storage architecture specifically for time series data. In their approach, they introduce an index structure for blockchains enabling an efficient access to the blocks and transactions in conjunction with a time series database for managing the time series data. The system decides for incoming queries whether they should be processed by the blockchain or the time series data and then forwards them accordingly. This approach reduces the overhead significantly, because on the one hand, time series databases are highly optimized to process time series queries. On the other hand, time series data are not immutable so that the data volume can be reduced as needed by deleting data that is no longer needed. However, this also represents the key weak point of this approach—the data in the time series databases are not protected against tampering or deletion.

Yet, there are research approaches towards tailored index structures especially for time series data in blockchains. Studies show that the performance of time series queries in blockchain systems can be increased significantly by such indices [125]. This could also improve the throughput of, for example, timestamp range queries (see Section 5).

Similar research approaches can be found for other specialized data and query types, such as index structures for location data in order to support efficient spatial queries, e.g., the work by Nurgaliev et al. [126].

Verifiable Queries and Database Systems. With verifiable queries, a user is able to verify the integrity of the result of a query. This ensures that the data and the execution have not been tampered with. For this purpose, a new class of database systems has emerged, the so-called verifiable database systems.

Zhang et al. [127] propose such a verifiable database system called *vSQL*, which supports arbitrary SQL queries. Here, a user is able to outsource a relational database to an untrusted server and has only to store a hash value locally. Then, the user can send SQL queries to that untrusted server and verify the integrity of the result. This verification is done by an interactive protocol, which utilizes interactive proofs.

Zhang et al. [128] propose another verifiable database system, which is called *Spitz*. It builds on top of *Forkbase* [129], which is a distributed multi-version storage engine utilizing the key-value data model, and maintains multiple index structures to facilitate verifiable query processing. The verification of the result of a query is done by comparing the hash value, which must be computed by using the proofs included in the result, with a previously locally stored hash value.

Zhou et al. [130] propose an SGX-based verifiable database system called *VeriDB*, which uses a trusted execution environment called *Intel SGX* (see https://www.intel.com/content/www/us/en/architecture-and-technology/software-guard-extensions.html; accessed on 15 December 2021) where data are isolated and encrypted in memory [131]. VeriDB hosts the query engine supporting SQL queries within an Intel SGX enclave, with the actual data residing in untrusted memory. The verification of the data is performed during the query processing by the query engine using a verifiable storage layer.

Table 5 summarizes the main findings regarding the characteristics and features of these six research directions in the field of query processing in blockchains.

Table 5. Summary of key findings regarding the state of the art.

Research Direction	Characteristics and Features
Blockchain Systems Enhanced with Database Features	A database layer is built on top of a blockchain system that is used as an integrity-protected data store. The database layer provides an interface for querying data efficiently. However, verifying the results of a query is an additional step, which increases the overhead significantly.
Database Systems Enhanced with Blockchain Features	A blockchain layer is built on top of a traditional database system. Data are queried directly from the database system. However, these database systems are not designed to detect tampered data during query processing.
Improvements to the Frontend Query Capabilities	Existing public blockchain systems such as Ethereum are internally modified or extended with a query layer to support familiar query languages such as SQL. However, queries regarding the data history are expensive.
Efficiency Improvements in Query Processing	Various techniques such as the parallelization of data processing or novel data structures enable more efficient querying of blockchain data. However, in order for query engines to benefit from this, they have first to be adapted accordingly.
Tailored Blockchain Optimizations for Specific Use Cases	Tailored index structures for blockchain systems increase the performance of specific types of queries such as time series queries. However, they are designed specifically for a certain use case, i.e., the blockchain system loses some of its universality.
Verifiable Queries and Database Systems	Verifiable queries are enabled over novel or existing database systems. However, these approaches do not necessarily require blockchain systems to be involved.

Blockchains were conceptually not developed to compete with traditional database systems in terms of data and query throughput. However, due to their inherent security features, they are increasingly used for managing important data. Considering the current state of technology, however, blockchains are still at the very beginning as far as query capabilities are concerned. Either one has to live with the native but naive query interfaces or the data processing takes place in a connected database system, which partially eliminates or at least reduces the security features. Therefore, there is a large body of research that aims to improve query capabilities in terms of usability, power, and performance. However, as our assessment of the state of research has shown, there are still many open research questions to be answered. In the following section, we elaborate on these open research questions.

7. Future Research Challenges

In this section, we elaborate on future research challenges based on the issues identified in Section 5 that need to be solved in order to enable an efficient query processing in blockchain systems. We group these issues into four categories of challenges, namely research challenges regarding *data models*, *data structures*, *block structures*, and *query processing*.

7.1. Data Models

In order to realize a JOIN operator for blockchains, a query engine has to be able to access, read, and process the common data stock of all involved blockchain systems. Since different blockchain systems have a highly heterogeneous technological infrastructure, a generic and standardized data model is needed that can be applied on all of these systems (see Issue A). Furthermore, for constant objects with object-dependent timestamps, it is useful to assign these timestamps a special status in the data model in order to access them more easily and in a standardized manner. To enable comparability of objects, it is worth considering introducing a type system, so that it is ensured that when comparing attributes of multiple objects with the same identifier, they are of the same type (see Issue B). Therefore, the first challenge is to create a standard for an expressive data model for blockchains. With such a data model, it must be possible to represent arbitrary kinds of data for any given use case. Triples, for example, have demonstrated their suitability in the context of RDF stores and could also be a beneficial approach for a blockchain data model.

7.2. Data Structures

In order to process queries on blockchain systems efficiently, state-of-the-art solutions operate a traditional database system in parallel to the actual blockchain. This database presents the current world state, i.e., the current value of the attributes of the objects stored in the blockchain. However, since these database systems cannot check the integrity of the data as required, an additional verification step is needed to check the results against the blockchain (see Issue C). To eliminate this verification step, it is necessary to come up with novel data structures, e.g., by combining search data structures with authenticated data structures such as *Merkle B-Trees* [132]. Such data structures are applied in current blockchain systems such as Ethereum. However, these structures are primarily used to facilitate the verification of transactions. A full-fledged support for comprehensive queries, as required by emerging use cases, is not provided by these structures. Therefore, the second challenge is to investigate how data structures can be designed that store generic data in a verifiable manner while providing fast access to the stored data.

7.3. Block Structures

There is some flexibility in organizing the data within a block. Data can either be physically clustered or added to useful data structures that allow efficient access to that data (see Issue D). It is also possible to construct index structures outside a block, but this would again require an additional verification step to check the results of a query against the blockchain. Thus, it is necessary to consider how the data are stored within a block. For example, different versions of the data can be stored within a block, each optimized for a certain type of query [133], similar to a triplestore with an *RDF3X engine* [134]. A lot of related work is concerned with the support of efficient spatio-temporal queries by adding special index structures to blockchain systems. Similar efforts are also needed for other types of data that are relevant in emerging application domains for blockchains (see Issue E). For example, expandable objects require special index structures in order to assemble them more efficiently. This can be done by storing pointers to their previous state, which simplifies left-folding. Similarly, constant objects also require index structures so that their history can be queried efficiently. Therefore, the third challenge is to investigate how the structure of a block could be designed to efficiently support different types of queries.

7.4. Query Processing

In query processing, the question is whether technically both object types (constant and expandable) can be supported at the same time (see Issue F). Even if this is technically possible, it could be contradictory from the perspective of the query language. The same question arises whether constant objects with object-dependent timestamps should be stored together with block-dependent timestamps (see Issue G). Another difficulty concerns the expandable objects, since their fields might be scattered over multiple blocks (see Issue H). During query processing, it is first necessary to locate the blocks that include the fields of the requested object, and then to assemble them by left-folding. Therefore, the fourth challenge is to investigate how query processing should be performed for each object type in order to efficiently compute the result of a query. Additionally, it is also necessary to investigate how a user can be supported in such a way that they can adequately formulate their queries.

The four research challenges mentioned above generally apply to any current blockchain system due to the conceptual design of blockchains. However, we expect that two factors will make these challenges even more difficult in the future, namely new blockchain architectures and legal restrictions.

7.5. New Blockchain Architectures

The fundamental architecture of a blockchain, as presented in Section 2, is constantly evolving. One trend that can be observed in this context is the so-called *sharding*. Sharding is introduced to address the typically low scalability of blockchains [135]. With blockchain sharding, the blockchain data is horizontally partitioned into shards where each shard is managed by a subset of the nodes in a network. One strategy in this regard can be to keep thematically related data in a common partition in order to create homogeneous partitions. A quite similar approach is known from traditional databases when a *snowflake schema* is applied. That is, data is divided among several tables in accordance with a specific dimension [136]. This makes queries regarding a certain topic highly efficient, since only a part of the data needs to be processed. However, the number of necessary joins increases if a comprehensive view on the entire data set is required. The same issue arises with sharding. As discussed in Issue A, blockchain systems are not designed to support joins efficiently. Moreover, the nodes that belong to an associated shard can only validate data they store. Therefore, when a join is made, the validation results from different shards must first be merged. For this reason, the data structures and block structures as well as the query processing must be adapted so that even complex JOIN operators can be executed efficiently.

Another emerging trend are the so-called *atomic cross-chain swaps*. Here, multiple parties exchange assets across multiple blockchains. Initially, this function was introduced so that different cryptocurrencies can be traded [137]. However, the exchanged assets are technically not limited to cryptocurrencies. That is, using cross-chain swaps, it is also possible to transfer data from one blockchain system to another [138]. Similar to sharding, this allows to create thematically homogeneous blockchains. Each blockchain provider would then only include data that corresponds to its respective topic. If necessary, external content can be imported from another blockchain via cross-chain swaps. Of course, this also results in the same challenges as with sharding, namely the high number of joins required to obtain a comprehensive view on the entire data set. Unlike sharding, where all partitions have at least the same technical foundation, cross-chain swap requires a wide variety of blockchain systems to interoperate in order to support cross-chain join operations. Thus, the data structures and block structures must also be created in a cross-blockchain manner.

7.6. Legal Restrictions

As illustrated in Section 3.1, blockchains are becoming increasingly popular for storing sensitive data, such as health data. However, such private data are protected by data protection laws, such as the *General Data Protection Regulation (GDPR)* [139]. Although blockchains are ideal for the secure storage and distribution of sensitive data in terms of immutability and tamper-resistance, they are fundamentally in conflict with data privacy principles [140]. Special categories of personal data, such as health data, however, are subject to a particularly high degree of protection—here data subjects must be granted full control over their data. To this end, comprehensive adjustments to a blockchain are necessary [141]. In particular, the *right to be forgotten* is in conflict with the immutability of a blockchain, and the *right to restriction of processing* contradicts the fully decentralized distribution of data to nodes that manage them autonomously. Moreover, it is impossible for data subjects to exercise their right to data minimization against individual data processors, since the data are tamper-resistant available in a blockchain [142].

However, such adjustments to make a blockchain GDPR-compliant also have a significant impact on query processing in blockchains. These implications concern two aspects in particular. On the one hand, due to the right to be forgotten DELETE statements are required. In the context of blockchains, however, this is technically difficult not only due to immutability, but also because of expandable objects. If such an object has to be deleted, initially all components of the object have to be found. These components can be distributed arbitrarily over all blocks of the blockchain. To support DELETE statements efficiently, data structures and block structures are required that exceed auxiliary structures found in current blockchains systems significantly. On the other hand, the access control in blockchain systems must be considerably refined in order to grant data subjects the legally guaranteed control over their data. Data subjects must be able to make fine-grained decisions about who should have access to which data. As a consequence, queries regarding the change history of objects become much more complex in particular. If a user has restricted access to some of the changes, only, it must be resolved how a history query can be executed in this case without having to process the restricted data. Expandable objects constitute a special challenge in this respect as well, since they can only be queried and assembled if all components can be accessed. If this cannot be guaranteed due to access restrictions, the data models and also the query processing itself have to be revised.

8. Conclusions

Blockchains are considered the new go-to technology in many application domains to store data in an immutable and tamper-resistant manner while ensuring high availability. A blockchain, however, is rather a conceptual design than a specific embodiment of a technology. Therefore, there are different implementations of a blockchain, each with their respective advantages and disadvantages. To support query capabilities on blockchain data, there are currently two prevalent approaches:

The first approach is to store all data in the blockchain and then execute the queries on it. The advantages of this approach are that the data history is fully available, and the data are protected by being immutable and tamper-resistant. The disadvantage of this approach is that query processing requires sequential traversal of the blocks, since there are no index structures to improve the efficiency of query processing.

The second approach is to operate a database in parallel to the blockchain. This database maintains the world state. This way, SQL-like queries can be executed efficiently, which is this approach's advantage. Its disadvantage is that such a database does not provide the data history. As a consequence, temporal queries and state-based queries are not or at least insufficiently supported. Furthermore, the authenticity of this data is not guaranteed by the blockchain. To this end, an additional verification step is required.

Therefore, to unlock the full potential of the blockchain technology (i.e., security and data history combined with comprehensive query capabilities), many research efforts are still needed (e.g., in terms of developing new index and data access structures for blockchains). In particular, we identified four categories of current research challenges in this regard: data models, data structures, block structures, and query processing.

In summary, the importance of blockchain systems as a secure data store is undeniable for a digitized society. However, there are still many research questions to be addressed before blockchains can compete with traditional database systems in terms of query capabilities and efficiency.

Author Contributions: Conceptualization, C.G., D.P. and C.S.; methodology, D.P. and C.S.; software, D.P.; validation, C.G., D.P. and C.S.; formal analysis, D.P. and C.S.; investigation, D.P. and C.S.; resources, B.M.; data curation, D.P. and C.S.; writing—original draft preparation, D.P. and C.S.; writing—review and editing, C.G., B.M., D.P. and C.S.; visualization, D.P. and C.S.; supervision, B.M. and C.S.; project administration, B.M. and D.P.; funding acquisition, B.M. All authors have read and agreed to the published version of the manuscript.

Funding: This research was funded by German Federal Ministry of Education and Research (BMBF) as part of the Software Campus program grant number 01IS17051.

Institutional Review Board Statement: Not applicable.

Informed Consent Statement: Not applicable.

Data Availability Statement: Not applicable.

Acknowledgments: We thank the anonymous reviewers for their valuable comments and suggestions which helped us to improve the content and presentation of this paper.

Conflicts of Interest: The authors declare no conflict of interest.

References

1. Faroukhi, A.Z.; El Alaoui, I.; Gahi, Y.; Amine, A. Big data monetization throughout Big Data Value Chain: A comprehensive review. *J. Big Data* **2020**, *7*, 3:1–3:22. [CrossRef]
2. Wiens, J.; Price, W.N.; Sjoding, M.W. Diagnosing bias in data-driven algorithms for healthcare. *Nat. Med.* **2020**, *26*, 25–26. [CrossRef]
3. Saetta, S.; Caldarelli, V. How to increase the sustainability of the agri-food supply chain through innovations in 4.0 perspective: A first case study analysis. *Procedia Manuf.* **2020**, *42*, 333–336. [CrossRef]
4. Ren, L.; Meng, Z.; Wang, X.; Zhang, L.; Yang, L.T. A Data-Driven Approach of Product Quality Prediction for Complex Production Systems. *IEEE Trans. Ind. Inform.* **2021**, *17*, 6457–6465. [CrossRef]
5. Stach, C.; Bräcker, J.; Eichler, R.; Giebler, C.; Mitschang, B. Demand-Driven Data Provisioning in Data Lakes: BARENTS—A Tailorable Data Preparation Zone. In Proceedings of the 23rd International Conference on Information Integration and Web Intelligence, iiWAS '21, Linz, Austria, 29 November–1 December 2021; pp. 191–202.
6. Diènea, B.; Rodrigues, J.J.; Diallo, O.; Ndoye, E.H.M.; Korotaev, V.V. Data management techniques for Internet of Things. *Mech. Syst. Signal Process.* **2020**, *138*, 106564:1–106564:19. [CrossRef]
7. Pavlou, K.E.; Snodgrass, R.T. Forensic Analysis of Database Tampering. *ACM Trans. Database Syst.* **2008**, *33*, 1–47. [CrossRef]
8. Iqbal, M.; Matulevičius, R. Blockchain as a Countermeasure Solution for Security Threats of Healthcare Applications. In Proceedings of the 19th Business Process Management Conference, BPM '21, Rome, Italy, 6–10 September 2021; pp. 67–84.
9. Chopade, R.; Pachghare, V. Data Tamper Detection from NoSQL Database in Forensic Environment. *J. Cyber Secur. Mobil.* **2021**, *10*, 421–450.
10. Nwosu, A.U.; Goyal, S.B.; Bedi, P. Blockchain Transforming Cyber-Attacks: Healthcare Industry. In Proceedings of the 11th International Conference on Innovations in Bio-Inspired Computing and Applications, IBICA '20, Online, 16–18 December 2020; pp. 258–266.
11. Maity, M.; Tolooie, A.; Sinha, A.K.; Tiwari, M.K. Stochastic batch dispersion model to optimize traceability and enhance transparency using Blockchain. *Comput. Ind. Eng.* **2021**, *154*, 107134:1–107134:12. [CrossRef]
12. Ge, C.; Liu, Z.; Fang, L. A blockchain based decentralized data security mechanism for the Internet of Things. *J. Parallel Distrib. Comput.* **2020**, *141*, 1–9. [CrossRef]
13. Nakamoto, S. Bitcoin: A Peer-to-Peer Electronic Cash System. White Paper, Bitcoin. 2008. Available online: https://klausnordby.com/bitcoin/Bitcoin_Whitepaper_Document_HD.pdf (accessed on 12 December 2021).
14. Zhu, Y.; Zhang, Z.; Jin, C.; Zhou, A.; Qin, G.; Yang, Y. Towards Rich Query Blockchain Database. In Proceedings of the 29th ACM International Conference on Information & Knowledge Management, CIKM '20, Ireland (Virtual Event), 19–23 October 2020; pp. 3497–3500.
15. Ruan, P.; Anh Dinh, T.T.; Lin, Q.; Zhang, M.; Chen, G.; Chin Ooi, B. Revealing Every Story of Data in Blockchain Systems. *ACM SIGMOD Rec.* **2020**, *49*, 70–77. [CrossRef]

16. Hellwig, D.; Karlic, G.; Huchzermeier, A. *Build Your Own Blockchain: A Practical Guide to Distributed Ledger Technology*; Management for Professionals; Springer Nature: Cham, Switzerland, 2020.

17. Krishnan, S.; Balas, V.E.; Golden Julie, E.; Robinson, Y.H.; Balaji, S.; Kumar, R. (Eds.) *Handbook of Research on Blockchain Technology*; Academic Press: London, UK; San Diego, CA, USA; Cambridge, UK; Oxford, UK, 2020.

18. Merkle, R.C. A Digital Signature Based on a Conventional Encryption Function. In Proceedings of the 7th Conference on Advances in Cryptology, CRYPTO '87, Santa Barbara, CA, USA, 16–20 August 1987; pp. 369–378.

19. Vujičić, D.; Jagodić, D.; Ranđić, S. Blockchain technology, bitcoin, and Ethereum: A brief overview. In Proceedings of the 2018 17th International Symposium INFOTEH-JAHORINA, INFOTEH '18, East Sarajevo, Bosnia and Herzegovina, 21–23 March 2018; pp. 1–6.

20. Yue, C.; Xie, Z.; Zhang, M.; Chen, G.; Ooi, B.C.; Wang, S.; Xiao, X. Analysis of Indexing Structures for Immutable Data. In Proceedings of the 2020 ACM SIGMOD International Conference on Management of Data, SIGMOD '20, Portland, OR, USA, 14–19 June 2020; pp. 925–935.

21. Ruoti, S.; Kaiser, B.; Yerukhimovich, A.; Clark, J.; Cunningham, R. Blockchain Technology: What is It Good For? *Commun. ACM* **2019**, *63*, 46–53. [CrossRef]

22. Nofer, M.; Gomber, P.; Hinz, O.; Schiereck, D. Blockchain. *Bus. Inf. Syst. Eng.* **2017**, *59*, 183–187. [CrossRef]

23. Muzammal, M.; Qu, Q.; Nasrulin, B. Renovating blockchain with distributed databases: An open source system. *Future Gener. Comput. Syst.* **2019**, *90*, 105–117. [CrossRef]

24. Dwork, C.; Naor, M. Pricing via Processing or Combatting Junk Mail. In Proceedings of the 12th Annual International Cryptology Conference, CRYPTO '92, Santa Barbara, CA, USA, 16–20 August 1992; pp. 139–147.

25. Chowdhury, M.J.M.; Colman, A.; Kabir, M.A.; Han, J.; Sarda, P. Blockchain Versus Database: A Critical Analysis. In Proceedings of the 2018 17th IEEE International Conference on Trust, Security and Privacy in Computing and Communications/12th IEEE International Conference on Big Data Science and Engineering, TrustCom/BigDataSE '18, New York, NY, USA, 1–3 August 2018; pp. 1348–1353.

26. Rehmani, M.H. Blockchain Technology and Database Management System. In *Blockchain Systems and Communication Networks: From Concepts to Implementation*; Springer International Publishing: Cham, Switzerland, 2021; Chapter 2, pp. 15–22.

27. Chen, S.; Zhang, J.; Shi, R.; Yan, J.; Ke, Q. A Comparative Testing on Performance of Blockchain and Relational Database: Foundation for Applying Smart Technology into Current Business Systems. In Proceedings of the 6th International Conference on Distributed, Ambient, and Pervasive Interactions, DAPI '18, Las Vegas, NV, USA, 15–20 July 2018; pp. 21–34.

28. Ozdayi, M.S.; Kantarcioglu, M.; Malin, B. Leveraging blockchain for immutable logging and querying across multiple sites. *BMC Med. Genom.* **2020**, *13*, 82–88. [CrossRef] [PubMed]

29. Eberhardt, J.; Tai, S. On or Off the Blockchain? Insights on Off-Chaining Computation and Data. In Proceedings of the 6th IFIP WG 2.14 European Conference on Service-Oriented and Cloud Computing, ESOCC '17, Oslo, Norway, 27–29 September 2017; pp 3–15,

30. Lo, S.K.; Xu, X.; Chiam, Y.K.; Lu, Q. Evaluating Suitability of Applying Blockchain. In Proceedings of the 2017 22nd International Conference on Engineering of Complex Computer Systems, ICECCS '17, Fukuoka, Japan, 5–8 November 2017; pp. 158–161.

31. Alladi, T.; Chamola, V.; Rodrigues, J.J.P.C.; Kozlov, S.A. Blockchain in Smart Grids: A Review on Different Use Cases. *Sensors* **2019**, *19*, 4862. [CrossRef]

32. Mollah, M.B.; Zhao, J.; Niyato, D.; Lam, K.Y.; Zhang, X.; Ghias, A.M.Y.M.; Koh, L.H.; Yang, L. Blockchain for Future Smart Grid: A Comprehensive Survey. *IEEE Internet Things J.* **2021**, *8*, 18–43. [CrossRef]

33. Gaber, T.; Ahmed, A.; Mostafa, A. PrivDRM: A Privacy-Preserving Secure Digital Right Management System. In Proceedings of the Evaluation and Assessment in Software Engineering, EASE '20, Trondheim, Norway, 15–17 April 2020; pp. 481–486.

34. Hei, Y.; Liu, J.; Feng, H.; Li, D.; Liu, Y.; Wu, Q. Making MA-ABE fully accountable: A blockchain-based approach for secure digital right management. *Comput. Netw.* **2021**, *191*, 108029:1–108029:12. [CrossRef]

35. Ren, Q.; Man, K.L.; Li, M.; Gao, B. Using Blockchain to Enhance and Optimize IoT-based Intelligent Traffic System. In Proceedings of the 2019 International Conference on Platform Technology and Service, PlatCon '19, Jeju, Korea, 28–30 January 2019; pp. 1–4.

36. Wang, Q.; Ji, T.; Guo, Y.; Yu, L.; Chen, X.; Li, P. TrafficChain: A Blockchain-Based Secure and Privacy-Preserving Traffic Map. *IEEE Access* **2020**, *8*, 60598–60612. [CrossRef]

37. Holderried, M.; Hoeper, A.; Holderried, F.; Heyne, N.; Nadalin, S.; Unger, O.; Ernst, C.; Guthoff, M. Attitude and potential benefits of modern information and communication technology use and telemedicine in cross-sectoral solid organ transplant care. *Sci. Rep.* **2021**, *11*, 9037:1–9037:9. [CrossRef]

38. Hörbst, A.; Ammenwerth, E. Electronic health records: A systematic review on quality requirements. *Methods Inf. Med.* **2010**, *49*, 320–336.

39. Lupton, D. *The Quantified Self*; Polity Press: Malden, MA, USA, 2016.

40. Heinemann, L. Continuous Glucose Monitoring (CGM) or Blood Glucose Monitoring (BGM): Interactions and Implications. *J. Diabetes Sci. Technol.* **2018**, *12*, 873–879. [CrossRef]

41. Isakadze, N.; Martin, S.S. How useful is the smartwatch ECG? *Trends Cardiovasc. Med.* **2020**, *30*, 442–448. [CrossRef]

42. Stach, C.; Steimle, F.; Franco da Silva, A.C. TIROL: The Extensible Interconnectivity Layer for mHealth Applications. In Proceedings of the 23rd International Conference on Information and Software Technologies, ICIST '17, Druskininkai, Lithuania, 12–14 October 2017; pp. 190–202.

43. Pham, H.L.; Tran, T.H.; Nakashima, Y. A Secure Remote Healthcare System for Hospital Using Blockchain Smart Contract. In Proceedings of the 2018 IEEE Globecom Workshops, GC Wkshps '18, Abu Dhabi, United Arab Emirates, 9–13 December 2018; pp. 1–6.

44. Spanakis, E.G.; Bonomi, S.; Sfakianakis, S.; Santucci, G.; Lenti, S.; Sorella, M.; Tanasache, F.D.; Palleschi, A.; Ciccotelli, C.; Sakkalis, V.; Magalini, S. Cyber-attacks and threats for healthcare—A multi-layer thread analysis. In Proceedings of the 2020 42nd Annual International Conference of the IEEE Engineering in Medicine Biology Society, EMBC '20, Montreal, QC, Canada, 20–24 July 2020; pp. 5705–5708.

45. Ball, M.J.; Smith, C.; Bakalar, R.S. Personal health records: Empowering consumers. *J. Healthc. Inf. Manag.* **2007**, *21*, 76–86.

46. Peng, Z.; Xu, C.; Wang, H.; Huang, J.; Xu, J.; Chu, X. P^2B-Trace: Privacy-Preserving Blockchain-Based Contact Tracing to Combat Pandemics. In Proceedings of the 2021 International Conference on Management of Data, SIGMOD/PODS '21, China (Virtual Event), 20–25 June 2021; pp. 2389–2393.

47. De Aguiar, E.J.; Faiçal, B.S.; Krishnamachari, B.; Ueyama, J. A Survey of Blockchain-Based Strategies for Healthcare. *ACM Comput. Surv.* **2020**, *53*, 27:1–27:27. [CrossRef]

48. Hasselgren, A.; Kralevska, K.; Gligoroski, D.; Pedersen, S.A.; Faxvaag, A. Blockchain in healthcare and health sciences—A scoping review. *Int. J. Med. Inform.* **2020**, *134*, 104040:1–104040:10. [CrossRef]

49. Khatoon, A. A Blockchain-Based Smart Contract System for Healthcare Management. *Electronics* **2020**, *9*, 94. [CrossRef]

50. Przytarski, D.; Stach, C.; Gritti, C.; Mitschang, B. A Blueprint for a Trustworthy Health Data Platform Encompassing IoT and Blockchain Technologies. In Proceedings of the ISCA 29th International Conference on Software Engineering and Data Engineering, SEDE '20, Las Vegas, NV, USA (Virtual Event), 19–21 October 2020; pp. 56–65.

51. Tanwar, S.; Parekh, K.; Evans, R. Blockchain-based electronic healthcare record system for healthcare 4.0 applications. *J. Inf. Secur. Appl.* **2020**, *50*, 102407:1–102407:13. [CrossRef]

52. Smith, F. The influence of Amatino Manucci and Luca Pacioli. *BSHM Bull. J. Br. Soc. Hist. Math.* **2008**, *23*, 143–156. [CrossRef]

53. Ellerman, D.P. The Mathematics of Double Entry Bookkeeping. *Math. Mag.* **1985**, *58*, 226–233. [CrossRef]

54. Brandon, D. The BLOCKCHAIN: The Future of Business Information Systems? *Int. J. Acad. Bus. World* **2016**, *10*, 33–40.

55. Beck, R.; Avital, M.; Rossi, M.; Thatcher, J.B. Blockchain Technology in Business and Information Systems Research. *Bus. Inf. Syst. Eng.* **2017**, *59*, 381–384. [CrossRef]

56. Carlin, T. Blockchain and the Journey Beyond Double Entry. *Aust. Account. Rev.* **2019**, *29*, 305–311. [CrossRef]

57. Faccia, A.; Moşteanu, N.R.; Leonardo, L.P. Blockchain Hash, the Missing Axis of the Accounts to Settle the Triple Entry Bookkeeping System. In Proceedings of the 2020 12th International Conference on Information Management and Engineering, ICIME '20, Amsterdam, The Netherlands, 16–18 September 2020; pp. 18–23.

58. Gökten, S.; Özdoğan, B. The Doors Are Opening for the New Pedigree: A Futuristic View for the Effects of Blockchain Technology on Accounting Applications. In *Digital Business Strategies in Blockchain Ecosystems: Transformational Design and Future of Global Business*; Hacioglu, U., Ed.; Springer International Publishing: Cham, Switzerland, 2020; pp. 425–438.

59. Schmitz, J.; Leoni, G. Accounting and Auditing at the Time of Blockchain Technology: A Research Agenda. *Aust. Account. Rev.* **2019**, *29*, 331–342. [CrossRef]

60. Sveistrup Søgaard, J. A blockchain-enabled platform for VAT settlement. *Int. J. Account. Inf. Syst.* **2021**, *40*, 100502:1–100502:18.

61. Zhang, Y.; Xiong, F.; Xie, Y.; Fan, X.; Gu, H. The Impact of Artificial Intelligence and Blockchain on the Accounting Profession. *IEEE Access* **2020**, *8*, 110461–110477. [CrossRef]

62. Femenia-Ribera, C.; Mora-Navarro, G.; Martinez-Llario, J.C. Advances in the Coordination between the Cadastre and Land Registry. *Land* **2021**, *10*, 81. [CrossRef]

63. Panda, S.K.; Mohammad, G.B.; Nandan Mohanty, S.; Sahoo, S. Smart contract-based land registry system to reduce frauds and time delay. *Secur. Priv.* **2021**, *4*, e172:1–e172:21. [CrossRef]

64. Zhang, P.; Schmidt, D.C.; White, J.; Lenz, G. Blockchain Technology Use Cases in Healthcare. In *Advances in Computers*; Raj, P., Deka, G.C., Eds.; Elsevier: Amsterdam, The Netherlands, 2018; Chapter 1; pp. 1–41.

65. Peiró, N.N.; Martinez García, E.J. Blockchain and Land Registration Systems. *Eur. Prop. Law J.* **2017**, *6*, 296–320. [CrossRef]

66. Vos, J. *Blockchain-Based Land Registry: Panacea, Illusion or Something in Between*; ELRA Annual Publication 7; European Land Registry Association: Brussels, Belgium, 2017.

67. Dabbagh, M.; Choo, K.K.R.; Beheshti, A.; Tahir, M.; Safa, N.S. A survey of empirical performance evaluation of permissioned blockchain platforms: Challenges and opportunities. *Comput. Secur.* **2021**, *100*, 102078:1–102078:13. [CrossRef]

68. Benarous, L.; Kadri, B.; Bouridane, A.; Benkhelifa, E. Blockchain-based forgery resilient vehicle registration system. In *Transactions on Emerging Telecommunications Technologies*; John Wiley & Sons: Hoboken, NI, USA, 2021; pp. 1–18.

69. Rosado, T.; Vasconcelos, A.; Correia, M. A Blockchain Use Case for Car Registration. In *Essentials of Blockchain Technology*; Li, K.C., Chen, X., Jiang, H., Bertino, E., Eds.; Chapman & Hall/CRC: New York, NY, USA, 2019; Chapter 10; pp. 205–234.

70. Sahai, A.; Pandey, R. Smart Contract Definition for Land Registry in Blockchain. In Proceedings of the 2020 IEEE 9th International Conference on Communication Systems and Network Technologies, CSNT '20, Gwalior, India, 2020, 10–12 April 2020; pp. 230–235.

71. Shinde, D.; Padekar, S.; Raut, S.; Wasay, A.; Sambhare, S.S. Land Registry Using Blockchain—A Survey of existing systems and proposing a feasible solution. In Proceedings of the 2019 5th International Conference On Computing, Communication, Control And Automation, ICCUBEA '19, Pune, India, 19–21 September 2019; pp. 1–6.

72. Singh Yadav, A.; Singh Kushwaha, D. Query Optimization in a Blockchain-Based Land Registry Management System. *Ingénierie Systèmes D'Inf.* **2021**, *26*, 13–21. [CrossRef]
73. Shah, R.; Meyer Goldstein, S.; Ward, P.T. Aligning supply chain management characteristics and interorganizational information system types: An exploratory study. *IEEE Trans. Eng. Manag.* **2002**, *49*, 282–292. [CrossRef]
74. Nastasijević, I.; Lakićević, B.; Petrović, Z. Cold chain management in meat storage, distribution and retail: A review. *IOP Conf. Ser. Earth Environ. Sci.* **2017**, *85*, 012022:1–012022:10. [CrossRef]
75. Tian, F. A supply chain traceability system for food safety based on HACCP, blockchain & Internet of things. In Proceedings of the 2017 International Conference on Service Systems and Service Management, ICSSSM '17, Dalian, China, 16–18 June 2017; pp. 1–6.
76. Stach, C.; Gritti, C.; Przytarski, D.; Mitschang, B. Trustworthy, Secure, and Privacy-aware Food Monitoring Enabled by Blockchains and the IoT. In Proceedings of the 2020 IEEE International Conference on Pervasive Computing and Communications Workshops, PerCom '20, Austin, TX, USA, 23–27 March 2020; pp. 50:1–50:4.
77. Fan, Y.; de Kleuver, C.; de Leeuw, S.; Behdani, B. Trading off cost, emission, and quality in cold chain design: A simulation approach. *Comput. Ind. Eng.* **2021**, *158*, 107442:1–107442:16. [CrossRef]
78. Menon, K.N.; Thomas, K.; Thomas, J.; Titus, D.J.; James, D. ColdBlocks: Quality Assurance in Cold Chain Networks Using Blockchain and IoT. In Proceedings of the 2nd International Conference on Emerging Technologies in Data Mining and Information Security, IEMIS '20, Kolkata, India, 2–4 July 2020; pp. 781–789.
79. Duan, J.; Zhang, C.; Gong, Y.; Brown, S.; Li, Z. A Content-Analysis Based Literature Review in Blockchain Adoption within Food Supply Chain. *Int. J. Environ. Res. Public Health* **2020**, *17*, 1784. [CrossRef]
80. Köhler, S.; Pizzol, M. Technology assessment of blockchain-based technologies in the food supply chain. *J. Clean. Prod.* **2020**, *269*, 122193:1–122193:10. [CrossRef]
81. Kayikci, Y.; Subramanian, N.; Dora, M.; Singh Bhatia, M. Food supply chain in the era of Industry 4.0: Blockchain technology implementation opportunities and impediments from the perspective of people, process, performance, and technology. In *Production Planning & Control*; Taylor & Francis: Zug, Switzerland; Saint Helier, Jersey, 2020; pp. 1–21.
82. Shahid, A.; Almogren, A.; Javaid, N.; Al-Zahrani, F.A.; Zuair, M.; Alam, M. Blockchain-Based Agri-Food Supply Chain: A Complete Solution. *IEEE Access* **2020**, *8*, 69230–69243. [CrossRef]
83. Zhang, X.; Sun, P.; Xu, J.; Wang, X.; Yu, J.; Zhao, Z.; Dong, Y. Blockchain-Based Safety Management System for the Grain Supply Chain. *IEEE Access* **2020**, *8*, 36398–36410. [CrossRef]
84. Gritzalis, D.A. Principles and requirements for a secure e-voting system. *Comput. Secur.* **2002**, *21*, 539–556. [CrossRef]
85. Gibson, J.P.; Krimmer, R.; Teague, V.; Pomares, J. A review of E-voting: The past, present and future. *Ann. Telecommun.* **2016**, *71*, 279–286. [CrossRef]
86. Boyd, C.; Gjøsteen, K.; Gritti, C.; Haines, T. A Blind Coupon Mechanism Enabling Veto Voting over Unreliable Networks. In Proceedings of the 20th International Conference on Cryptology in India, INDOCRYPT '19, Hyderabad, India, 15–18 December 2019; pp. 250–270.
87. Haines, T.; Gritti, C. Improvements in Everlasting Privacy: Efficient and Secure Zero Knowledge Proofs. In Proceedings of the 4th International Joint Conference on Electronic Voting, E-Vote-ID '19, Bregenz, Austria, 1–4 October 2019; pp. 116–133.
88. Moura, T.; Gomes, A. Blockchain Voting and Its Effects on Election Transparency and Voter Confidence. In Proceedings of the 18th Annual International Conference on Digital Government Research, dg.o '17, Staten Island, NY, USA, 7–9 June 2017; pp. 574–575.
89. Wani, S.; Imthiyas, M.; Almohamedh, H.; Alhamed, K.M.; Almotairi, S.; Gulzar, Y. Distributed Denial of Service (DDoS) Mitigation Using Blockchain–A Comprehensive Insight. *Symmetry* **2021**, *13*, 227. [CrossRef]
90. Hanifatunnisa, R.; Rahardjo, B. Blockchain based e-voting recording system design. In Proceedings of the 2017 11th International Conference on Telecommunication Systems Services and Applications, TSSA '17, Lombok, Indonesia, 26–27 October 2017; pp. 1–6.
91. Hjálmarsson, F.Þ.; Hreiðarsson, G.K.; Hamdaqa, M.; Hjálmtýsson, G. Blockchain-Based E-Voting System. In Proceedings of the 2018 IEEE 11th International Conference on Cloud Computing, CLOUD '18, San Francisco, CA, USA, 2–7 July 2018; pp. 983–986.
92. Kshetri, N.; Voas, J. Blockchain-Enabled E-Voting. *IEEE Softw.* **2018**, *35*, 95–99. [CrossRef]
93. Ruparel, H.; Hosatti, S.; Shirole, M.; Bhirud, S. Secure Voting for Democratic Elections: A Blockchain-Based Approach. In Proceedings of the 2020 International Conference on Communication, Computing and Electronics Systems, ICCCES '20, Coimbatore, India, 21–22 October 2020; pp. 615–628.
94. Wang, B.; Sun, J.; He, Y.; Pang, D.; Lu, N. Large-scale Election Based On Blockchain. *Procedia Comput. Sci.* **2018**, *129*, 234–237. [CrossRef]
95. Buchmann, A.; Koldehofe, B. Complex Event Processing. *IT-Inf. Technol.* **2009**, *51*, 241–242. [CrossRef]
96. Pratama, F.A.; Mutijarsa, K. Query Support for Data Processing and Analysis on Ethereum Blockchain. In Proceedings of the 2018 International Symposium on Electronics and Smart Devices, ISESD '18, Bandung, Indonesia, 23–24 October 2018; pp. 1–5.
97. Zhu, Y.; Zhang, Z.; Jin, C.; Zhou, A.; Yan, Y. SEBDB: Semantics Empowered BlockChain DataBase. In Proceedings of the 2019 IEEE 35th International Conference on Data Engineering, ICDE '19, Macao, China, 8–11 April 2019; pp. 1820–1831.
98. Heuer, A.; Scholl, M.H. Principles of Object-Oriented Query Languages. In *Datenbanksysteme in Büro, Technik und Wissenschaft*; Appelrath, H.J., Ed.; Springer: Berlin/Heidelberg, Germany, 1991; pp. 178–197.
99. Libkin, L. Expressive power of SQL. *Theor. Comput. Sci.* **2003**, *296*, 379–404. [CrossRef]

100. Klyne, G.; Newman, C. *Date and Time on the Internet: Timestamps*; Standards Track RFC 3339, July; IETF—Network Working Group: Reston, VA, USA; Geneva, Switzerland, 2002.
101. Androulaki, E.; Barger, A.; Bortnikov, V.; Cachin, C.; Christidis, K.; De Caro, A.; Enyeart, D.; Ferris, C.; Laventman, G.; Manevich, Y.; et al. Hyperledger Fabric: A Distributed Operating System for Permissioned Blockchains. In Proceedings of the Thirteenth EuroSys Conference, EuroSys '18, Porto, Portugal, 23–26 April 2018; pp. 30:1–30:15.
102. Ruan, P.; Dinh, T.T.A.; Loghin, D.; Zhang, M.; Chen, G.; Lin, Q.; Ooi, B.C. Blockchains vs. Distributed Databases: Dichotomy and Fusion. In Proceedings of the 2021 International Conference on Management of Data, SIGMOD/PODS '21, China (Virtual Event), 20–25 June 2021; pp. 1504–1517.
103. El-Hindi, M.; Binnig, C.; Arasu, A.; Kossmann, D.; Ramamurthy, R. BlockchainDB: A Shared Database on Blockchains. *Proc. VLDB Endow.* **2019**, *12*, 1597–1609. [CrossRef]
104. Peng, Y.; Du, M.; Li, F.; Cheng, R.; Song, D. FalconDB: Blockchain-Based Collaborative Database. In Proceedings of the 2020 ACM SIGMOD International Conference on Management of Data, SIGMOD '20, Portland, OR, USA, 14–19 June 2020; pp. 637–652.
105. Allen, L.; Antonopoulos, P.; Arasu, A.; Gehrke, J.; Hammer, J.; Hunter, J.; Kaushik, R.; Kossmann, D.; Lee, J.; Ramamurthy, R.; Setty, S.; Szymaszek, J.; van Renen, A.; Venkatesan, R. Veritas: Shared Verifiable Databases and Tables in the Cloud. In Proceedings of the 9th Biennial Conference on Innovative Data Systems Research, CIDR '19, Asilomar, CA, USA, 13–16 January 2019; pp. 111:1–111:9.
106. BigchainDB GmbH. *BigchainDB 2.0: The Blockchain Database*; White Paper; BigchainDB GmbH: Berlin, Germany, 2018.
107. Nathan, S.; Govindarajan, C.; Saraf, A.; Sethi, M.; Jayachandran, P. Blockchain Meets Database: Design and Implementation of a Blockchain Relational Database. *Proc. VLDB Endow.* **2019**, *12*, 1539–1552. [CrossRef]
108. Schuhknecht, F.M.; Sharma, A.; Dittrich, J.; Agrawal, D. chainifyDB: How to get rid of your Blockchain and use your DBMS instead. In Proceedings of the 11th Annual Conference on Innovative Data Systems Research, CIDR '21, online, 11–15 January 2021; pp. 4:1–4:10.
109. Eisenberg, A.; Melton, J. SQL: 1999, Formerly Known as SQL3. *ACM SIGMOD Rec.* **1999**, *28*, 131–138. [CrossRef]
110. Wood, G. Ethereum: A Secure Decentralised Generalised Transaction Ledger. Ethereum Yellow Paper Berlin Version 888949c, Ethereum Project. 2021. Available online: https://files.gitter.im/ethereum/yellowpaper/VIyt/Paper.pdf (accessed on 12 December 2021).
111. Han, J.; Kim, H.; Eom, H.; Coignard, J.; Wu, K.; Son, Y. Enabling SQL-Query Processing for Ethereum-based Blockchain Systems. In Proceedings of the 9th International Conference on Web Intelligence, Mining and Semantics, WIMS '19, Seoul, Korea, 26–28 June 2019; pp. 9:1–9:7.
112. Tong, X.; Tang, H.; Jiang, N.; Fan, W.; Gao, Y.; Deng, S.; Zhang, Z.; Jin, C.; Yang, Y.; Qin, G. SQL-Middleware: Enabling the Blockchain with SQL. In Proceedings of the 26th International Conference on Database Systems for Advanced Applications, DASFAA '21, Taipei, Taiwan, 11–14 April 2021; pp. 622–626.
113. Li, Y.; Zheng, K.; Yan, Y.; Liu, Q.; Zhou, X. EtherQL: A Query Layer for Blockchain System. In Proceedings of the 22nd International Conference on Database Systems for Advanced Applications, DASFAA '17, Suzhou, China, 27–30 March 2017; pp. 556–567.
114. Bragagnolo, S.; Marra, M.; Polito, G.; Gonzalez Boix, E. Towards Scalable Blockchain Analysis. In Proceedings of the 2019 IEEE/ACM 2nd International Workshop on Emerging Trends in Software Engineering for Blockchain, WETSEB '19, Montreal, QC, Canada, 27 May 2019; pp. 1–7.
115. Xu, C.; Zhang, C.; Xu, J. vChain: Enabling Verifiable Boolean Range Queries over Blockchain Databases. In Proceedings of the 2019 International Conference on Management of Data, SIGMOD '19, Amsterdam, The Netherlands, 30 June–5 July 2019; pp. 141–158.
116. Xing, X.; Chen, Y.; Li, T.; Xin, Y.; Sun, H. A blockchain index structure based on subchain query. *J. Cloud Comput.* **2021**, *10*, 52:1–52:11. [CrossRef]
117. Jia, D.Y.; Xin, J.C.; Wang, Z.Q.; Lei, H.; Wang, G.R. SE-Chain: A Scalable Storage and Efficient Retrieval Model for Blockchain. *J. Comput. Sci. Technol.* **2021**, *36*, 693–706. [CrossRef]
118. Peng, Z.; Wu, H.; Xiao, B.; Guo, S. VQL: Providing Query Efficiency and Data Authenticity in Blockchain Systems. In Proceedings of the 2019 IEEE 35th International Conference on Data Engineering Workshops, ICDEW 19, Macao, China, 8–12 April 2019; pp. 1–6.
119. Wu, H.; Peng, Z.; Guo, S.; Yang, Y.; Xiao, B. VQL: Efficient and Verifiable Cloud Query Services for Blockchain Systems. *IEEE Trans. Parallel Distrib. Syst.* **2022**, *33*, 1393–1406. [CrossRef]
120. Gritti, C.; Önen, M.; Molva, R. Privacy-Preserving Delegable Authentication in the Internet of Things. In Proceedings of the 34th ACM/SIGAPP Symposium on Applied Computing, SAC '19, Limassol, Cyprus, 8–12 April 2019; pp. 861–869.
121. Stach, C.; Bräcker, J.; Eichler, R.; Giebler, C.; Gritti, C. How to Provide High-Utility Time Series Data in a Privacy-Aware Manner: A VAULT to Manage Time Series Data. *Int. J. Adv. Secur.* **2020**, *13*, 88–108.
122. Wortner, P.; Schubotz, M.; Breitinger, C.; Leible, S.; Gipp, B. Securing the Integrity of Time Series Data in Open Science Projects using Blockchain-based Trusted Timestamping. In Proceedings of the Workshop on Web Archiving and Digital Libraries held in conjunction with the 18th ACM/IEEE Joint Conference on Digital Libraries, WADL '19, Champaign, IL, USA, 2 June 2019; pp. 2:1–2:3.

123. Dhanush, G.A.; Raj, K.S.; Kumar, P. Blockchain Aided Predictive Time Series Analysis in Supply Chain System. In Proceedings of the 2021 2nd International Conference on Electrical and Electronics Engineering, ICEEE '21, NCR New Delhi, India, 2–3 January 2021; pp. 913–925.
124. Yu, Z.; Cai, Y.; Hong, W. A Storage Architecture of Blockchain for Time-Series Data. In Proceedings of the 2019 2nd International Conference on Hot Information-Centric Networking, HotICN '19, Chongqing, China, 13–15 December 2019; pp. 90–91.
125. Qu, Q.; Nurgaliev, I.; Muzammal, M.; Jensen, C.S.; Fan, J. On spatio-temporal blockchain query processing. *Future Gener. Comput. Syst.* **2019**, *98*, 208–218. [CrossRef]
126. Nurgaliev, I.; Muzammal, M.; Qu, Q. Enabling Blockchain for Efficient Spatio-Temporal Query Processing. In Proceedings of the 19th International Conference on Web Information Systems Engineering, WISE '18, Dubai, United Arab Emirates, 12–15 November 2018; pp. 36–51.
127. Zhang, Y.; Genkin, D.; Katz, J.; Papadopoulos, D.; Papamanthou, C. vSQL: Verifying Arbitrary SQL Queries over Dynamic Outsourced Databases. In Proceedings of the 2017 IEEE Symposium on Security and Privacy, SP '17, San Jose, CA, USA, 22–26 May 2017; pp. 863–880.
128. Zhang, M.; Xie, Z.; Yue, C.; Zhong, Z. Spitz: A Verifiable Database System. *Proc. VLDB Endow.* **2020**, *13*, 3449–3460. [CrossRef]
129. Wang, S.; Dinh, T.T.A.; Lin, Q.; Xie, Z.; Zhang, M.; Cai, Q.; Chen, G.; Ooi, B.C.; Ruan, P. Forkbase: An Efficient Storage Engine for Blockchain and Forkable Applications. *Proc. VLDB Endow.* **2018**, *11*, 1137–1150. [CrossRef]
130. Zhou, W.; Cai, Y.; Peng, Y.; Wang, S.; Ma, K.; Li, F. VeriDB: An SGX-Based Verifiable Database. In Proceedings of the 2021 International Conference on Management of Data, SIGMOD/PODS '21, China (Virtual Event), 20–25 June 2021; pp. 2182–2194.
131. McKeen, F.; Alexandrovich, I.; Anati, I.; Caspi, D.; Johnson, S.; Leslie-Hurd, R.; Rozas, C. Intel® Software Guard Extensions (Intel® SGX) Support for Dynamic Memory Management Inside an Enclave. In Proceedings of the Hardware and Architectural Support for Security and Privacy 2016, HASP '16, Seoul, Korea, 18 June 2016; pp. 1–9.
132. Li, F.; Hadjieleftheriou, M.; Kollios, G.; Reyzin, L. Dynamic Authenticated Index Structures for Outsourced Databases. In Proceedings of the 2006 ACM SIGMOD International Conference on Management of Data, SIGMOD '06, Chicago, IL, USA, 27–29 June 2006; pp. 121–132.
133. Przytarski, D. Using Triples as the Data Model for Blockchain Systems. In Proceedings of the Blockchain enabled Semantic Web Workshop and Contextualized Knowledge Graphs Workshop co-located with the 18th International Semantic Web Conference, BlockSW/CKG@ISWC '19, Auckland, New Zealand, 26–30 October 2019; pp. 1–2.
134. Neumann, T.; Weikum, G. The RDF-3X engine for scalable management of RDF data. *VLDB J.* **2010**, *19*, 91–113. [CrossRef]
135. Dang, H.; Dinh, T.T.A.; Loghin, D.; Chang, E.C.; Lin, Q.; Ooi, B.C. Towards Scaling Blockchain Systems via Sharding. In Proceedings of the 2019 International Conference on Management of Data, SIGMOD '19, Amsterdam, The Netherlands, 30 June–5 July 2019; pp. 123–140.
136. Levene, M.; Loizou, G. Why is the snowflake schema a good data warehouse design? *Inf. Syst.* **2003**, *28*, 225–240. [CrossRef]
137. Herlihy, M. Atomic Cross-Chain Swaps. In Proceedings of the 2018 ACM Symposium on Principles of Distributed Computing, PODC '18, Egham, UK, 23–27 July 2018; pp. 245–254.
138. Xu, J.; Ackerer, D.; Dubovitskaya, A. A Game-Theoretic Analysis of Cross-Chain Atomic Swaps with HTLCs. In Proceedings of the 2021 IEEE 41st International Conference on Distributed Computing Systems, ICDCS '21, Washington, DC, USA, 7–10 July 2021; pp. 584–594.
139. European Parliament and Council of the European Union. Regulation on the protection of natural persons with regard to the processing of personal data and on the free movement of such data, and repealing Directive 95/46/EC (Data Protection Directive). Legislative Acts L119. *Off. J. Eur. Union* **2016**.
140. Hofman, D.; Lemieux, V.L.; Joo, A.; Alves Batista, D. "The margin between the edge of the world and infinite possibility": Blockchain, GDPR and information governances. *Rec. Manag. J.* **2019**, *29*, 240–257. [CrossRef]
141. Shi, S.; He, D.; Li, L.; Kumar, N.; Khan, M.K.; Choo, K.K.R. Applications of blockchain in ensuring the security and privacy of electronic health record systems: A survey. *Comput. Secur.* **2020**, *97*, 101966:1–101966:20. [CrossRef] [PubMed]
142. Tatar, U.; Gokce, Y.; Nussbaum, B. Law versus technology: Blockchain, GDPR, and tough tradeoffs. *Comput. Law Secur. Rev.* **2020**, *38*, 105454:1–105454:11. [CrossRef]

future internet

Article

Enable Fair Proof-of-Work (PoW) Consensus for Blockchains in IoT by Miner Twins (MinT)

Qian Qu [1], Ronghua Xu [1], Yu Chen [1,*], Erik Blasch [2] and Alexander Aved [2]

[1] Department of Electrical and Computer Engineering, Binghamton University, Binghamton, NY 13902, USA; qqu2@binghamton.edu (Q.Q.); rxu22@binghamton.edu (R.X.)

[2] The U.S. Air Force Research Laboratory, Rome, NY 13441, USA; erik.blasch.1@us.af.mil (E.B.); alexander.aved@us.af.mil (A.A.)

* Correspondence: ychen@binghamton.edu

Abstract: Blockchain technology has been recognized as a promising solution to enhance the security and privacy of Internet of Things (IoT) and Edge Computing scenarios. Taking advantage of the Proof-of-Work (PoW) consensus protocol, which solves a computation intensive hashing puzzle, Blockchain ensures the security of the system by establishing a digital ledger. However, the computation intensive PoW favors members possessing more computing power. In the IoT paradigm, fairness in the highly heterogeneous network edge environments must consider devices with various constraints on computation power. Inspired by the advanced features of Digital Twins (DT), an emerging concept that mirrors the lifespan and operational characteristics of physical objects, we propose a novel Miner Twins (MinT) architecture to enable a fair PoW consensus mechanism for blockchains in IoT environments. MinT adopts an edge-fog-cloud hierarchy. All physical miners of the blockchain are deployed as microservices on distributed edge devices, while fog/cloud servers maintain digital twins that periodically update miners' running status. By timely monitoring of a miner's footprint that is mirrored by twins, a lightweight Singular Spectrum Analysis (SSA)-based detection achieves the identification of individual misbehaved miners that violate fair mining. Moreover, we also design a novel Proof-of-Behavior (PoB) consensus algorithm to detect dishonest miners that collude to control a fair mining network. A preliminary study is conducted on a proof-of-concept prototype implementation, and experimental evaluation shows the feasibility and effectiveness of the proposed MinT scheme under a distributed byzantine network environment.

Keywords: digital twin; blockchain; Proof-of-Work; microservices; Singular Spectrum Analysis (SSA); byzantine fault tolerance

Citation: Qu, Q.; Xu, R.; Chen, Y.; Blasch, E.; Aved, A. Enable Fair Proof-of-Work (PoW) Consensus for Blockchains in IoT by Miner Twins (MinT). *Future Internet* **2021**, *13*, 291. https://doi.org/10.3390/fi13110291

Academic Editor: Christoph Stach

Received: 30 October 2021
Accepted: 16 November 2021
Published: 19 November 2021

Publisher's Note: MDPI stays neutral with regard to jurisdictional claims in published maps and institutional affiliations.

1. Introduction

Advancement in Internet of Things (IoT), Edge Computing, Big Data (BD), and Artificial Intelligence (AI)/Machine Learning (ML) technologies makes the concept of Smart Cities realistic. However, widely adopting IoT-based applications and services in smart cities also brings new security and privacy concerns. Thanks to multiple attractive features including decentralization, auditability and traceability, blockchain has been widely recognized as a great potential to revolutionize the fundamentals of information and communication technology (ICT) [1]. Applying blockchain to smart cities is promising to bring efficiency, scalability and security properties to IoT-based applications, such as smart surveillance [2], privacy preservation [3], decentralized data marketplaces [4], time banking of community [5], identity authentication [6] and access control [7,8].

Digital Twins (DT) is being developed to optimize manufacturing and aviation processes [9]. By monitoring, simulating and mirroring the status of a physical object (PO), DT can build an intelligent and evolving system model based on the logic object (LO). Leveraging data fusion and AI/ML algorithms, DT can be used to predict the behavior of the PO given some specific situations or environments. Similar to DT, the Dynamic Data

Driven Applications Systems (DDDAS) concept developed in the late 1990s seeks to use modeling to support predictive expectations based on the coordination with models and data [10]. Thus, DDDAS can determine optimized solutions or even failure preventive actions on POs to enable an intelligent and resilient system.

Research has been conducted to apply blockchain to enable many attractive features in DTs, including transparency, decentralization, data immutability and Peer-to-Peer (P2P) communication [11]. However, directly integrating existing blockchain technologies into the highly heterogeneous IoT environments presents critical challenges in terms of scalability, performance, security and fairness [12]. Some permissioned blockchains use a Practical Byzantine Fault Tolerance (PBFT) [13] protocol, which demonstrates high throughput and low latency but only allows for a very limited network scalability in terms of the number of validators. Most permissionless blockchain networks utilize a hashing-intensive Proof-of-Work (PoW) consensus protocol to achieve security and scalability guarantees. Due to the various computation capability of miners, mining centralization in a PoW blockchain not only leads to inequity of rewarding among participants but also brings about security issues, such as majority (51%) attacks [14].

Inspired by the essential features of DTs, mirroring and monitoring, this paper proposes a novel edge-fog-cloud Miner Twins (MinT) architecture to enable a fair PoW consensus mechanism for blockchains in IoT environments. In the MinT architecture, the fog/cloud sever establishes and maintains digital twins for the miners of the blockchain, which are deployed as microservices in edge devices that participate in the blockchain network. Container technology is adopted to encapsulate PoW algorithm as microservices, and each containerized miner is dedicated to mining tasks using pre-configured computation power. As each miner has the same constrained computation resources, it becomes affordable to optimize resource limited IoT devices.

In summary, this paper makes the following contributions:

(1) A secure-by-design MinT architecture is introduced to allow for fair-mining-as-a-service (FMaaS) in heterogeneous IoT environments;
(2) We propose a novel miner twin-enabled fair-mining mechanism, which can monitor the computing resources usage at miners and can regularly apply anomaly detection to deter misbehaved nodes from unfairly overwhelming honest peers using extra computing power;
(3) A lightweight SSA Singular Spectrum Analysis (SSA)-based detection is designed to identify individual misbehaved miners that violate fair mining policies, while a Proof-of-Behavior consensus algorithm is designed to detect multiple Byzantine miners that collude to compromise a fair mining network; and
(4) A proof-of-concept prototype is implemented and tested on a small-scale private PoW mining network, and experimental results verified that the MinT is feasible and effective to ensure a fair mining system.

The remainder of this paper is organized as follows: Section 2 reviews the background on blockchain and PoW consensus and then briefly discusses the state-of-the-art research on DT. Section 3 introduces the rationale and architecture of MinT. The miner twin-enabled fair-mining mechanism including SSA and PoB-based detection algorithms is explained in Section 4.1. Section 5 presents the prototype implementation with numerical results. Section 6 concludes the paper with future work.

2. Related Work

This section introduces blockchain and PoW consensus background knowledge. Following that, we describe digital twin technology and how DT can be used to guarantee the fair mining scheme in blockchain.

2.1. Blockchain and Nakamoto Consensus Protocol

As a form of distributed ledger technology (DLT), *Blockchain* was initially implemented as an enabling technology of Bitcoin [15], which aimed to provide a cryptocurrency to record and verify commercial transactions among trustless entities in a decentralized manner. With the decentralized P2P network architecture and cryptographic mechanisms, participants in a blockchain system maintain the immutability and auditability of data and transactions recorded on the distributed ledger instead of relying on a centralized third party trust authority.

As one of the most fundamental problems in a distributed/decentralized computing environment, *consensus* in a blockchain network can be defined as a fault-tolerant state-machine replication problem, which aims to maintain the globally distributed ledger state across the P2P network. Bitcoin adopts the Nakamoto consensus based on a Proof-of-Work (PoW) scheme to achieve pseudonymity, scalability and probabilistic finality in an asynchronous and open-access network environment. The goal of Nakamoto consensus is to ensure all participants agree on a common network transaction log as a serialized blockchain [12].

PoW is essentially an incentive-based consensus algorithm, which requires all participants to compete for rewards through a cryptographic block discovery racing game. To be a winner in PoW block generation, every miner has to solve a computing-intensive hash puzzle problem. In brief, a valid PoW solution requires exhaustively querying a cryptographic hash function for a partial preimage generated from a candidate block [16]. Finally, the hash code of a candidate block must satisfy a predefined difficulty condition parameter h, such as having a fixed length of bits as zeros.

Given current *block_data*, which consists of a block header and ordered transactions by time stamps, a miner continually calculates a hash value *nonce* until it satisfies the PoW puzzle problem. The PoW puzzle problem can be formally defined as follows:

$$hash\_block = \mathcal{H}(block\_data|nonce) \leqslant D(h), \tag{1}$$

where for some fixed length of bits L and difficulty condition, $D(h) = 2^{L-h}$. $\mathcal{H}(\cdot)$ is a predefined collision-resistant cryptographic hash function that outputs a hash string $L \in \{0,1\}^{\lambda}$, and λ is the length of a hash string.

The PoW process defined by Equation (1) is essentially a verifiable process of a weighted random coin-tossing [12]. Thus, the probability of generating a valid block is in proportion to miners' computation resources. Higher computation power leads to higher hash string rate in PoW, which means more rewards and benefits. Such a mining centralization may discourage participants who have limited computation resources, such as IoT devices; but it also lead to majority (51%) attacks if an adversary controls more than 50% of the computation resources of the whole network.

To reduce energy consumption in PoW consensus, Peercoin [17] proposed Proof-of-Stake (PoS), which requires a miner to use its coin stake to solve the puzzle solution. Unlike PoW protocols that relies on a brute-force hash calculation, PoS miners use a process of "virtual mining" manner that only consumes limited computational resources. However, PoS still has a mining centralization issue because an attacker can amplify its power by simply accumulating the credit stake. As the first practical Byzantine Fault Tolerant (BFT) consensus, Practical BFT (PBFT) [13] guarantees both liveness and safety in synchronous network environments given the assumption that at most of $\lfloor \frac{n-1}{3} \rfloor$ out of total of n participants in consensus protocol are Byzantine faults. As PBFT requires that all nodes communicate synchronously to achieve consensus purposes, it has poor scalability due to high latency and communication overhead as more nodes join the consensus network.

2.2. Digital Twins

The concept of DT was proposed in 2002 and archived in a NASA white paper in 2014 [18]. Essentially, a DT is a digital representation of the components and dynamics of a physical system [19]. Based on the functionalities, DTs can be roughly categorized into three kinds: monitoring DTs, simulational DTs and operational DTs [20]. As suggested by the names, monitoring twins allow system operators to monitor the status of a physical system; simulation twins can predict the future status of the physical system in different scenarios using various simulation tools and ML algorithms; and operational twins is a *complex sensing and control system* that enabled human operators to interact with a cyber-physical system and to perform different actions in addition to monitoring, analysis and prediction [21], which is similar to human–machine teaming [22].

Earlier studies on DT mainly focused on the area of manufacturing covering different key factors for smart manufacturing including simulation, optimization and the use of AI. For instance, an event-driven simulation for manufacturing and assembly tasks based on Digital Twin and human–robot collaboration was presented [23]. A DT-based framework was proposed to achieve high precision and multidisciplinary coupling during the assembly process, which mainly focused on High precision products (HPPs) workshops [24]. HPP also establishes a predict and optimization model as well as a case study to verify the effectiveness and feasibility. A case study presented an ice cream machine as an application example of DT in food industry [25], which focused on the visualization and interaction based on virtual reality (VR) and augmented reality (AR) technologies. Secure data transmission was also highlighted in the framework by employing a secure gate between machine and cloud.

Recently, efforts are reported in variant aspects of smart cities including Smart Driving, Smart Grid and Smart Healthcare. For instance, the optimization issue in the electric propulsion drive systems (EPDS) of self-driving electric vehicles were discussed [26]. In the proposed DT-based framework, the connection between a logical twin in the control software with the propulsion motor drive system enables EPDS performance estimation. However, there were no experimental results presented after giving the concepts of the platform. A behaviors-based algorithm was proposed to help the drivers avoid potential risk [27]. Combining the ML techniques and DT relies on the connectivity of the system and faces challenges in optimization and accuracy [28]. A case study has been reported that tackles the management of wind farm using DT and cloud technologies combined with big data analysis to build remote control station [29].

Recently, some healthcare applications redefined DT by including living objects [30]. A DT-based healthcare framework was proposed for monitoring and predicting the health condition of an individual using wearable devices [31]. A DT-based remote surgery prototype was introduced consisting of VR, 4G and AI to create a digital twin of a patient and to realize real-time surgery over mobile network [32]. Due to the fast development of telecommunication technologies, 5G and beyond networks are very complicated as they are expected to support more emerging applications with more diverse requirements [33]. The community is considering DT as an efficient, cost-effective approach to accelerate the design, test and implementation of 5G/6G networks [34].

Due to the foreseeable importance and popularity of DT in IoT, 5G/6G and edge computing, blockchain is adopted to enhance the security, trust and reliability of DTs [11,35]. The work reported in this paper, however, is the first in this area that leverages DT to tackle the unfair mining problem in the PoW consensus protocol. Using digital twins, MinT monitors the computing resource utility of the miners and quickly detects abusers using Singular Spectrum Analysis (SSA) [36], one of the fastest change point detection algorithms [37]. Our MinT also uses a Proof-of-Behavior (PoB) consensus algorithm to guarantee byzantine tolerant anomaly detection.

3. MinT: Rationale and Architecture

Aiming at a secure-by-design fair PoW mining network in heterogeneous IoT environments, our MinT scheme leverages DT technology to continuously monitor the usage of containerized miners and discourages misbehaving nodes from unfairly overwhelming the peers by using extra computing power. Figure 1 illustrates the high-level system architecture of MinT, which adopts a hierarchical cloud-fog-edge computing paradigm. Such a hierarchical framework not only provides system scalability for large-scale fair mining tasks based on geographically distributed IoT devices but also supports flexible management and coordinated central and decentralized local decisions given heterogeneous networks and application domains. Moreover, MinT relies on a permissioned network that provides basic security guarantees, such as the public key infrastructure (PKI) and digital signature, data integrity [2], identity authentication [6] and access control [38], etc. In essential, MinT is a partial decentralized PoW mining network. Furthermore, DTs in MinT are mainly used to monitor their associated miners and to support misbehavior detection, and they do not directly participate in the PoW mining task or impose interference on the consensus protocol. Therefore, our Mint is promising in enabling a fair mining network without sacrificing distribution and decentralization. The rationale behind the MinT is described as follows:

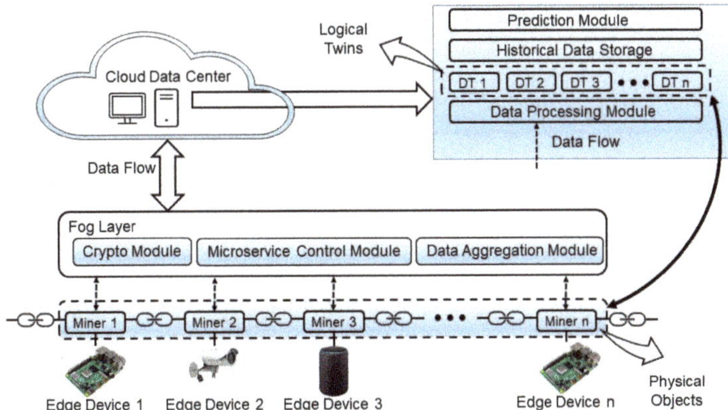

Figure 1. Illustration of MinT system architecture.

1. Containerized PoW Miner: The edge layer in MinT consists of various types of IoT devices, such as smart cameras in a surveillance system or smart meters connected to a power grid. To follow an ideal "one cup-one vote" Nakamoto consensus protocol, the Pow algorithm is encapsulated into containers as physical miners that are deployed on edge devices to participate in the blockchain network, and all containers are assigned the same computation resource for PoW mining process. Each miner has the same probability of generating blocks and being rewarded accordingly due to the uniform computation distribution of the network. Thus, these containerized PoW miners construct a fair mining blockchain network disregarding devices' capability.

2. Microservice-oriented Service: MinT utilizes an intermediate fog layer to provide middle-ware services for devices at edge and cloud level. To address heterogeneity of IoT systems, a lightweight Microservice-oriented architecture (MoA) is adopted as a fundamental service infrastructure to support functionality, such as data aggregation and microservice management, and security mechanisms, such as encryption/decryption; to identity verification; to access control, etc. Each microservice unit exposes a set of RESTful web-service APIs for interaction. The fine-granularity and loose-coupling features of the MoA framework allow for fast development and easy deployment among heterogeneous platforms using non-standard development.

3. DT-enabled Fair Mining Intelligence: As dishonest containerized miners could use extra computing power than they are permitted, MinT relies on DT technology and intelligent services on a fog/cloud server to maintain a fair mining network at the edge layer. By aggregating data flows from distributed physical miners, mirroring miners (logic objects) that are associated with their physical counterparts are created and managed by the fog or cloud server. These miner twins monitor the usage of containerized miners running on devices. By analyzing the real-time status of miner twins and historical statistics, abusers can be detected and preventive actions can be triggered to deter identified misbehaving miners such that the MinT ensures a fair mining blockchain network.

4. Miner Twin-Enabled Fair-Mining Mechanism

This section provides a comprehensive overview of the MinT-based fair mining mechanism such that readers can understand the key components and workflow. Then, we describe the miner twin process including key parameter selection. Following that, we offer details on lightweight SSA-based anomaly detection and the byzantine tolerant PoB consensus algorithm.

4.1. MinT Workflow for Fair Mining

Figure 2 illustrates the workflow of the fair-mining mechanism in the MinT system. The upstream data flow starts from the containerized miners and aggregates the fog servers installed with different modules. The fog server first normalize the data from all physical miners, which reports to it under its jurisdiction. The fog server can either construct logical miners that mirror these new physical miners or update the status of existing logical miners. The fog server further encrypts its local logical twining miners and forwards them to the cloud.

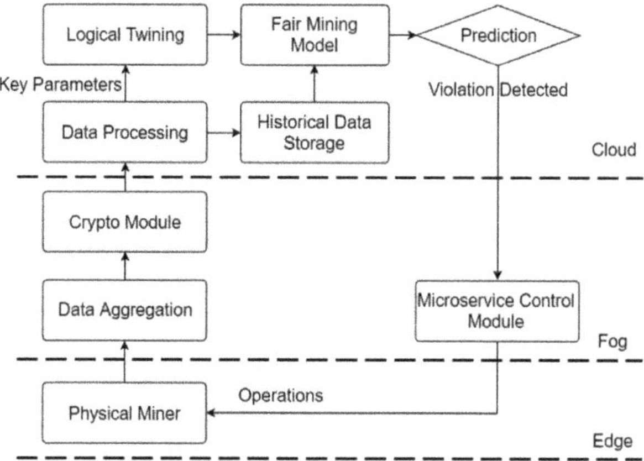

Figure 2. Miner twin-based fair-mining flowchart.

Upon receiving the encrypted data from multiple fog servers, the cloud server aggregates the information into a logical miners pool to represent a system level twinning PoW network. Using the live feed from the logical twin and the historical data, MinT uses an intelligent model for fair mining strategy. Given a fair mining algorithm, the upstream data flow starts from the predication. The predicted status is compared with the actual footprint, using anomaly detection algorithm MinT; identifies dishonest miners who violate the fair PoW consensus; and sends orders to the Microservice Control Module on a fog layer accordingly, which takes further actions on the "outlaws".

4.2. Miner Twin Process

The notations used in this paper are listed in Table 1. To mirror the physical miner, several parameters are extracted for the logical miner, including central CPU usage (C), global GPU usage (G), memory usage (M) and I/O bandwidth (B). Since PoW depends on computation intensive algorithms, the CPU usage and GPU usage are chosen as the key parameters according to the selection of calculation module, while memory, I/O bandwidth and other metrics are considered as contributing parameters. To avoid falling behind other miners, the physical miner normally uses all of the allocated CPU/GPU resources.

As the system resource allocated to each miner is restricted but identical, the data can be normalized in the form of percentages, for example $c = \frac{C}{C_{set}} \times 100\%$, where C_{set} is the preset CPU limit and c is the normalized value. Given an assumption that a containerized miner can only use its CPU to perform the PoW algorithm, then for a miner k, the parameter vector of its Physical Object (physical miner) with timestamp i would be $PO_{ki} = (c_{ki}, g_{ki}, m_{ki}, b_{ki})$, and the Key Parameter is c_{ki}. The vector for the Logical Object (logic miner) can be represented as $LO_{ki} = (c_{ki}, g_{ki}, m_{ki}, b_{ki})$, and the Key Parameter is c_{ki}.

Table 1. Relevant basic notations.

Symbol	Descriptions	Symbol	Descriptions
PO_{ki}	parameter vector of miner	LO_{ki}	parameter vector of twin
c_{ki}	cpu usage	g_{ki}	gpu usage
m_{ki}	memory usage	b_{ki}	I/O bandwidth
$\mathbb{X}$	target time series	X	trajectory matrix
N	target series length	L	SSA window length
$\vec{X}_i$	lagged vectors	K	numbers of lagged vectors
λ_L	eigenvalues	U	left singular matrix
V	right singular matrix	I	subset indices
X_I	reconstructed matrix	$\mathbb{X}_{\mathbb{I}}$	reconstructed time series
X_{test}	test matrix	$\vec{X}_j$	vectors of test matrix
p	starting point of test matrix	q	ending point of test matrix
Q	window length of test matrix	$D_{n,I,p,q}$	sum of the squared distances
S_n	normalized sum	W_n	CUSUM of squared distances
$\mu_{n,I}$	estimator	κ	constant of W_n
h	threshold for W_n	t_α	quantile of the standard normal distribution
$\mathcal{N}$	mining network	n_i	miners
m_i	dishonest miners	f	fraction of dishonest
B_i	behavior vector	G	global view of B_i
B^*	benchmark of B_i	$s(i)$	consensus score
s^*	ground truth of $s(i)$	d	POB window length

4.3. Fast Anomaly Detection for Fair Mining

Fast and accurate identification of the misbehaved miners is an essential step to ensuring fair mining, where MinT adopts the Singular Spectrum Analysis (SSA) algorithm to achieve this goal. SSA is recognized as one of the quickest sequential change-point detection approaches for processing time series problems [39]. By decomposing and reconstructing the interested time series, SSA extracts certain components of the origin series such as periodic pattern, noises, trends, etc. SSA is widely used in solving problems such as smoothing, extraction of seasonality components, as well as study the structure in some minor time series and change-point detection [36].

Unlike traditional methods, SSA is non-parametric and does not require prior knowledge of the parametric model of the considered time series data. Although SSA uses some statistical concepts, it does not need any statistical assumptions about the target series. Moreover, SSA algorithm can be used for processing time series with relatively small size, which make this method more suitable for edge-fog scenarios [40]. The SSA algorithm can be divided into four steps as (see Moskvina et al. at 2003) [41]:

1. Embedding: The target of SSA is a one-dimensional time series $\mathbb{X} = [x_1, \ldots, x_N]$, where N is the series length. By choosing proper window length L, one can transfer the times series into multi-dimensional series of vectors $\vec{X}_i$. Combine these vectors results in the trajectory matrix $X = [\vec{X}_1, \vec{X}_2, \ldots, \vec{X}_K]$, where $K = N - L + 1$. The multi-dimensional vectors $\vec{X}_i = (x_i, \ldots, x_{L+i-1})', i = 1, \ldots, K$, are also called lagged vectors.

2. Singular Value Decomposition (SVD) [42]: After singular value decomposing the trajectory matrix X, the eigenvalues are denoted by $\lambda_1, \ldots, \lambda_L$ in decreasing order of magnitude and the corresponding eigenvectors $U_1, \ldots, U_L$ where the matrix $U = [U_1, U_2 \ldots, U_L]$ and $\|U_i\| = 1$ is orthogonal. Then, the eigentriples are $(\sqrt{\lambda_i}, U_i, V_i)$, by denoting $V_i = X' U_i / \sqrt{\lambda_i}$. Supposing that the rank of X is d, then the trajectory matrix is $X = X_1 + \ldots + X_d$.

3. Grouping and Reconstructing: The next step is to group the matrices X_i into certain groups and to calculate the sum within these groups. Therefore, we denote a subset indices $I = i_1, i_2, \ldots, i_l$, where $l < L$. Therefore, the corresponding matrix is $X_I = X_{i_1} + \ldots + X_{i_l}$.

4. Diagonal Averaging: Using diagonal averaging, we can transfer X_I into time series $\mathbb{X}_I$.

$$\mathbb{X}_I(i) = \begin{cases} \frac{1}{i} \sum_{j=1}^{i} x_{j,i-j+1} & \text{for} \quad 1 \leq i < L \\ \frac{1}{L} \sum_{j=1}^{L} x_{j,i-j+1} & \text{for} \quad L \leq i \leq K \\ \frac{1}{N-i+1} \sum_{j=i-K+1}^{N-K+1} x_{j,i-j+1} & \text{for} \quad K \leq i \leq N. \end{cases} \tag{2}$$

By selecting certain subset indices $I = i_1, i_2, \ldots, i_l$, one can reconstruct the time series. By observing the distance between the l-dimensional matrix and the test time series matrix, we can detect the anomaly by identifying a significant increase in the distance. The SSA-based Change-Point detection utilized in the paper can be described in following stages [41]:

Stage 1: Construct Base Matrix First, construct the base matrix (or target matrix) according to the four steps of the SSA algorithm. Given the target time series $\mathbb{X} = [x_{n+1}, \ldots, x_{n+N}]$, embed it into the trajectory matrix $X = [\vec{X}_1, \vec{X}_2, \ldots, \vec{X}_K]$, where $K = N - L + 1$. Then, the columns of the trajectory matrix are the vectors:

$$\vec{X}_i = (x_{n+i}, \ldots, x_{n+L+i-1})', i = 1, \ldots, K. \tag{3}$$

Then, conduct the SVD and get L eigenvectors which can be grouped into certain subset $I = i_1, i_2, \ldots, i_l, l < L$.

Stage 2: Construct Test Matrix Similarly, we select integers p, q and Q where $Q = q - p > 0$. Then, we construct the test matrix of size $L \times Q$:

$$X_{test} = [\vec{X}_{p+1}, \vec{X}_{p+2}, \ldots, \vec{X}_{p+Q}], \tag{4}$$

and the columns of the matrix are the vectors:

$$\vec{X}_j = (x_{n+j}, \ldots, x_{n+L+j-1})', j = p+1, \ldots, p+Q, \tag{5}$$

Stage 3: Compute the Detection Statistics In this stage, we first compute $D_{n,l,p,q}$, the sum of the squared Euclidean distances between the l-dimensional subspace from the base matrix and the vectors $\vec{X}_j (j = p+1, \ldots, p+Q)$ from the test matrix.

$$D_{n,l,p,q} = \sum_{j=p+1}^{q} ((\vec{X}_j)^T \vec{X}_j - (\vec{X}_j)^T U U^T \vec{X}_j). \tag{6}$$

Then, we give the normalized sum of squared distances

$$S_n = \frac{1}{\mu_{n,l}} \tilde{D}_{n,l,p,q}, \tag{7}$$

where $\tilde{D}_{n,l,p,q} = \frac{1}{LQ} D_{n,l,p,q}$ and $\mu_{n,l} = \tilde{D}_{m,l,0,K}$ is the estimator and we make the hypothesis that no change of time series structure occurs at the time intervals where m is the largest value of $m \leq n$.

We also compute the Cumulative Sum (CUSUM) W_n of the normalized sum of squared distances as the final score for the anomaly detection.

$$W_1 = S_1, W_{n+1} = max\{0, W_n + S_{n+1} - S_n - \kappa / \sqrt{LQ}\}, n \geq 1, \tag{8}$$

where κ is a constant, and in this paper, we set $\kappa = 1/(3\sqrt{LQ})$ [43].

Stage 4: Set threshold and make decisions To detect the change of the time series, we could check the values of $D_{n,l,p,q}$, S_n and W_n. Basicallys the large value of the three detection statistics indicates the change or the anomaly. In this paper, we choose the W_n-based detection algorithm as it gives greater sensitivity compared with the former two detection statistics [41]. The algorithm announces a structural change if we observe $W_n > h$ for some n where h is the threshold given by

$$h = \frac{2t_\alpha}{LQ}\sqrt{\frac{1}{3}Q(3LQ - Q^2 + 1)}, \tag{9}$$

and t_α is the $1 - \alpha$-quantile of the standard normal distribution [41].

4.4. Proof-of-Behavior Consensus Algorithm for Fair Mining Enforcement

The abovementioned SSA-based detection can identify a single misbehaved miner based on its own footprint; however, it cannot handle byzantine scenarios that multiple compromised miners by an adversary collude to violate fair mining policies. By observing a miner's running operations, the calculated cumulative sum (CUSUM)-type W can indicate a miner's behavior. Inspired by deepfake detection in video surveillance systems [44,45], our MinT relies on a novel *Proof-of-Behavior* consensus algorithm that leverages CUSUM-type W calculated in SSA algorithm to detect multiple dishonest miners in distributed byzantine tolerant scenarios.

We consider a mining network $\mathcal{N}$ including n_i miners, where $i \in \{1, k\}$ and $k = |\mathcal{N}|$. All dishonest miners are denoted by $m_i \in \mathcal{M}$ and their fraction is $f = |\mathcal{M}|/|\mathcal{N}|$. We use observed CUSUM-type W_i of miner n_i to demote a behavior vector $B_i = \{b_1, b_2, \ldots, b_d\}$, where $b_k = w_k \in W_i$ and d is the SSA detection time window. Finally, each twin can maintain a global view of collected behavior vectors, which is a matrix $G = \{B_1, B_2, \ldots, B_k\}$. The PoB firstly generates a behavior score $s(i)$ for each miner n_i, which is a sum of relative Euclidean distances between other miners' behavior vector. Then, a $B_i \in G$ with minimal behavior score is selected as a benchmark B^*.

The PoB consensus algorithm aims to chooses a behavior vector B, which deviates at least from the distribution of G. However, an adversary can compromise multiple miners that generate large vectors to force "honest" miners to choose a byzantine behavior vector as the ground truth one. Thus, our PoB algorithm adopts a *Krum* aggregation rule to guarantee byzantine tolerance. We assume that honest miners within network $\mathcal{N}$ store G including $n \geq 2f + 3$ vectors in which at most f vectors are generated by byzantine nodes in $\mathcal{M}$. For B_j belongs to the $n - f - 2$ closest vectors to B_i, where $i \neq j$, we denote $i \rightarrow j$. Therefore, we could define the consensus score:

$$s(i) = \sum_{i \rightarrow j} ||B_i - B_j||^2. \tag{10}$$

Then, each node can compute behavior scores $s(1), \ldots, s(k)$ that are associated with miners $n_1, \ldots, n_k$ separately. By calculating the minimum behavior score

$$s^* = \min_{i \in \{1, \ldots, k\}} (s(i)), \tag{11}$$

all honest miners choose a behavior vector B_i that satisfies $s(i) = s^*$ as the ground truth B^*. Given assumption that an adversary controls no more than f miners, all honest miners can reach an agreement on the unique B^*.

5. Experimental Study

In this section, a proof-of-concept prototype implementation and experimental configuration are described. Following that, we evaluate effectiveness of the proposed MinT solution based on numerical results. Finally, we discuss performance and security properties provided by MinT.

5.1. Experimental Setup

A proof-of-concept test platform is created, in which 16 Raspberry Pis (RPi) are adopted as the edge devices. Each RPi is empowered with quad-core Cortex-A72 CPU @1.5GHz and an installed RAM with 4GB memory running Raspbian OS based on Debian. The single-board computer (SBC) is capable of carrying containerized PoW module to participate the blockchain network. A desktop functions as a fog server, which has Intel Core i7-7700K CPU and a RAM of 16 GB memory. All of the RPis are connected to a fog server via local area network (LAN).

As the GPU is not available on the RPi, we select a CPU-based PoW algorithm for container construction. For fast deployment, Docker [46] is adopted as the microservice container that is affordable to RPis and transmits the data from the physical miner to a fog server through RESTfull APIs. Each of the miner containers is configured with and restricted to one CPU core, 500 MB memory and 10 percent of system I/O bandwidth. The collected data are stored in forms of vector as described in Section 4.2.

As the PoW algorithm is executed on CPU, samples of the key parameter C are collected and the historical data vector c_{hi} is used to obtain the statistic profile, where $h = 1, \ldots, 16$ and $i = 0, 1, \ldots$. For SSA based change-point detection, as the standard SSA recommendation in the book [47], we define $N = 24$ according to the size of the data sets, $L = 12$ to the half size of N, $p = 12$, $q = 24$ and $d = 1s$. We deliberately set $p \geq K$ so that the base and test matrices would not coincide. After visual inspection of the components of the decomposition of the whole time series, we choose certain l to represent ignoring the noise components. To guarantee the accuracy and reliability, we repeat each experiment scenario for at least five times and over two hours each time to avoid contingency.

5.2. Experimental Results

All 16 miners, by default, run at 100% of the assigned system resources under the jurisdiction of the fog server. Four different test scenarios are considered in our experimental study. To verify SSA-based detection on a single misbehaved miner, we first conduct test cases that only one dishonest miner uses double-assigned computation power on mining given a different parameter combination. Then, we consider a more stealthy single miner violation, which incrementally increases the computing power from 20% up to 50%. To validate effectiveness of PoB-based detection, we simulate a byzantine network, in which two miners act as byzantine nodes while 14 miners are honest members. Finally, we evaluate the false-positive rates at the network level with different threshold settings.

5.2.1. SSA-Based Detection on Static Single Miner Violation

In this scenario, one dishonest miner uses twice as much CPU power as the assigned amount at $t = 200$ s. Figure 3a presents the network level observation at the fog server. The blue line is the average CPU usage for all 16 miners in this blockchain network, and the red line is the w_n value calculated using SSA algorithm as the score. The green line is the threshold $h = 0.607$, which is computed with $t_\alpha = 1.2815$. As shown by Figure 3a, the fluctuation in the average CPU utility incurs a low peak in the distance score. However, applying the SSA algorithm on each miner twin individually avoids the false negative. Figure 3b shows that a significant peak is observed at $t = 200$ s.

We also studied the impacts of different selections of the SSA parameters varying l and $q - p$ combination. Figure 3c shows the consequence of increasing the value of l from 4 to 8 but with the same matrix size. The larger l leads to a more noise part with the signal; therefore, it would be more difficult to find a change in the signal time series. If the l is too small, which would cause underfitting, we might miss some part of the signal. Due to limited space, the figure is not included here.

Meanwhile, the matrix size $q - p$ also has significant impact on the detection distance score. Figure 3d shows that, by increasing the value of $q - p$ to 24 while $l = 4$, the distance (red) line is smoother than in Figure 3b.

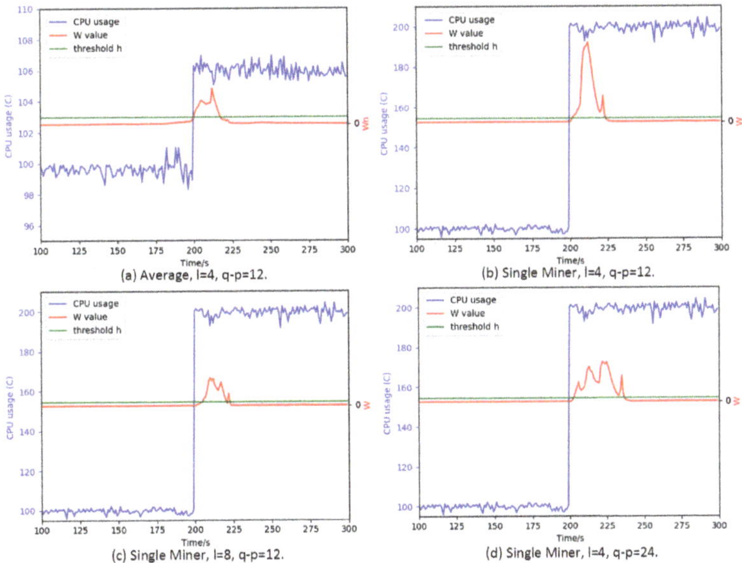

Figure 3. SSA detection on single miner violation with different parameter combinations. (**a**) Network level observation at the fog server; (**b**) Observation on a single miner; (**c**) Impacts of increasing l from 4 to 8 with the same matrix size; (**d**) Impacts of increasing the matrix size to 24 while maintain $l = 4$.

5.2.2. SSA Detection on Adaptive Single Miner Violation

The second scenario considers more stealthy behavior of a violator, which increases the computing power slowly, from 20% to 50%, taking multiple steps at time point $t = 125$ s, $t = 175$ s, $t = 225$ s and $t = 275$ s. Figure 4a shows the detection results in which a miner increases 20% CPU usage at each time point. Figures 4b–d show similar results of cases when the CPU usage increases by 30%, 40% and 50% respectively. Obviously, the SSA-based anomaly detection is able to detect the changes in the structure of the time series data and to identify the corresponding violation on mining power. However, the critical issue is how to select a threshold to ensure a high detection accuracy and to minimize the false-positive/negative rates.

5.2.3. PoB-Based Fair Mining Detection Effectiveness

We take an observation of 20 min on the 16 miners running at 100% of the assigned system resource. Two of the miners act as the byzantine (dishonest) workers, which would gain extra 10% at the 9th and 10th min. As shown in Figure 5a, the behavior vector B from dishonest workers varies from honest ones when the byzantine workers gain more computing power. During the two minutes where violation occurs, the resulting consensus scores associated with the byzantine nodes are much larger, as shown in Figure 5b.

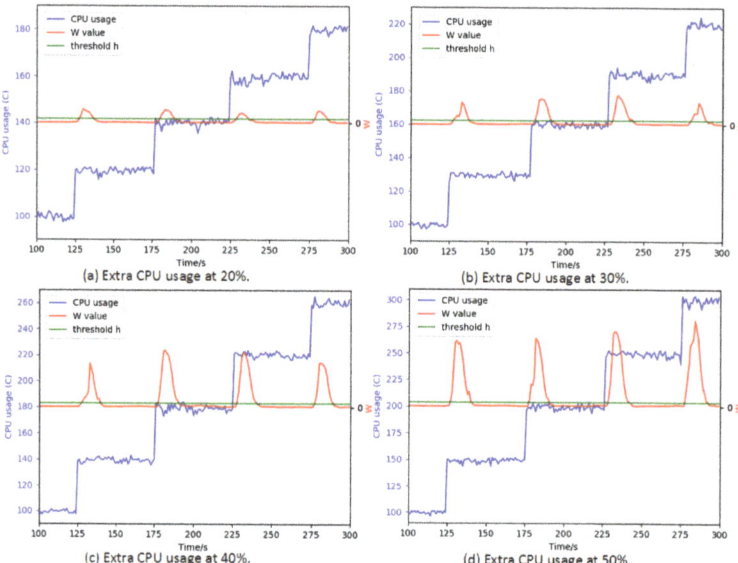

Figure 4. SSA detection on single miner violation with additive CPU usage. (**a**) One single miner increases 20% CPU usage at each time point; (**b**) One single miner increases 30% CPU usage at each time point; (**c**) One single miner increases 40% CPU usage at each time point; (**d**) One single miner increases 50% CPU usage at each time point.

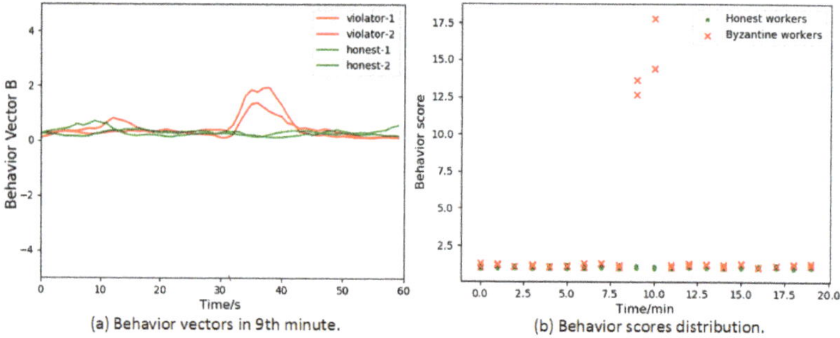

Figure 5. Behavior score distribution with sequential time spots. (**a**) The behavior vector from dishonest workers (red) varies from honest ones (green) when the byzantine workers gain more computing power; (**b**) Comparison between consensus scores associated with the byzantine nodes (red) and the honest nodes (green).

5.2.4. Fair Mining Violation Detection Performance Analysis

The fourth scenario is designed to mainly test the false positive rate from the network level observation at the fog server with different threshold settings. Figure 6 shows the false alarm rates when two of the sixteen miners gain extra system resources from 10% to 80%. The false alarm rate is calculated by comparing the averaged the W value with the threshold h. When we decrease h from 0.6 to 0.03, the false alarm rate increases rapidly at the beginning and then slowly approaches one. With the increasing percentage of the computing power the dishonest miner gains, the false alarm rate grows.

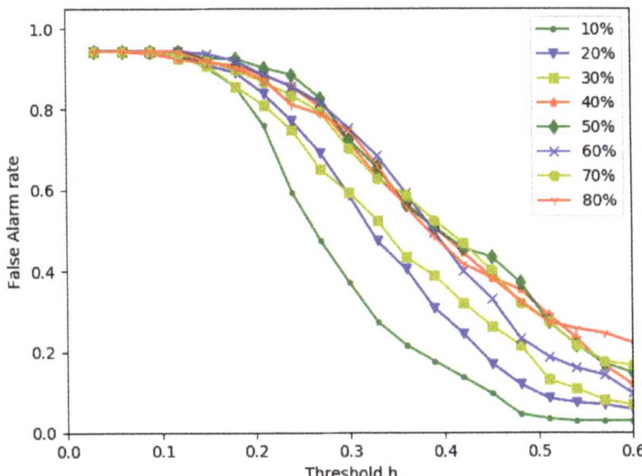

Figure 6. False alarm rate with different threshold h.

5.3. Discussions

The experimental results presented in this section are merely a preliminary study on top of a proof-of-concept platform. Our MinT relies on a permissioned network to provide basic security primitives, such as identity registration, authentication and access control, etc. Given the assumption that the adversary cannot control microservice control the module to send false parameters, we verify that SSA-based detection can identify a single dishonest miner that uses either static or adaptive mining violation strategies. Regarding byzantine scenarios that multiple dishonest miners collude to disturb fair mining mechanism, the PoB consensus algorithm adopts Krum rule in behavior score calculation, which only chooses $n - f - 2$ closet behavior vectors and precludes those $f - 1$ malicious vectors that are far away from the center of distribution. Given the assumption that an adversary cannot control more than f nodes of a mining network $\mathcal{N}$ that satisfies $n \geq 2f + 3$, all honest participants can still make agreements and output the unique benchmark behavior vector B^*.

Our MinT architecture envisions large-scale IoT networks based on a hierarchy of edge-fog-cloud paradigm. However, there are open questions that need to be addressed before bringing the proposed framework into real-world applications. We leave them for our future work.

- Although experimental results verify feasibility of SSA-based fair-mining violation detection, there still need investigation on SSA performance and accuracy given the impact of parameters, such as optimal/sub-optimal threshold selection and detection latency as scaling up miners.
- The PoB consensus is promising to guarantee byzantine fault tolerance in mining violation detection; however, the threat model based on attack scenarios in SSA detection needs more investigation, such as communication security between miner and twin and container's robustness given failed or compromised conditions. Therefore, the security mechanisms for communication between PO and LO, and container management are among the tasks of top priority.
- It is inevitable that extra overheads are incurred by security enforcement and data synchronization in fair-mining mechanism. Therefore, a comprehensive performance evaluation of the twinning process is necessary, such as computation and communication cost, network latency and storage requirement, etc.
- Furthermore, we also need to tackle scalability and heterogeneity issues such as as applying MinT into large-scale IoT networks. A hierarchical federated network frame-

work is promising to handle the trilemma in blockchain solutions that decentralization, security and scalability cannot perfectly co-exist [4].

6. Conclusions and Future Work

In this paper, we proposed MinT, an edge-fog-cloud architecture to enable a fair PoW consensus mechanism by leveraging miner twins. Experimentally, the paper validated the feasibility of the concept of using DT to monitor the miners' behaviors and to deter selfish nodes who violate the fair-mining rule. The reported preliminary results verify the effectiveness of using quick change point detection and the PoB consensus algorithm to catch fair mining violators; however, more intelligent solutions are needed to support dynamicity and optimization in fair mining network. Moreover, the above mentioned open questions need to be addressed in IoT-based mining networks. Our future work includes the following.

- We will conduct a comprehensive evaluation on SSA method in anomaly detection, especially for detection accuracy and performance, and the impact of parameter selection. Moreover, AI/ML-based algorithms will be investigated to improve anomaly detection accuracy and to support efficient dynamic resources management in the fair mining network.
- To apply MinT in a large-scale application scenario such as a smart surveillance system [48], we will implement a fully function prototype based on edge-fog-cloud architecture, in which physical containerized miners are on edge devices while digital twins are in the fog or cloud. Then, we will make a comprehensive performance analysis and assessment of security features.
- Furthermore, MinT relies on microservices that encapsulate a fair PoW mining algorithm into independent containers running on host machines. Thus, the security and privacy of containers and data reliability are among the top concerns. We will investigate the security of the container running environment, and data audition and integrity in microservice-to-microservice communication.

Author Contributions: Conceptualization, Q.Q., R.X. and Y.C.; methodology, Q.Q. and R.X.; software, Q.Q. and R.X.; validation, Q.Q., R.X. and Y.C.; formal analysis, Q.Q. and R.X.; investigation, Y.C.; resources, Y.C., E.B. and A.A.; data creation, Q.Q.; writing—original draft preparation, Q.Q. and R.X.; writing—review and editing, Y.C., E.B. and A.A.; visualization, Q.Q. and R.X.; supervision, Y.C. and E.B.; project administration, Y.C. and A.A.; funding acquisition, Y.C. All authors have read and agreed to the published version of the manuscript.

Funding: This work is partially supported by the U.S. National Science Foundation (NSF) via grants CNS-2141468.

Data Availability Statement: Not Applicable, the study does not report any data.

Acknowledgments: The views and conclusions contained herein are those of the authors and should not be interpreted as necessarily representing the official policies or endorsements, either expressed or implied, of the U.S. Air Force.

Conflicts of Interest: The authors declare no conflicts of interest.

Abbreviations

The following abbreviations are used in this manuscript:

AI	Artificial Intelligence
AR	Augmented Reality
BFT	Byzantine Fault Tolerant
CUSUM	Cumulative Sum
DDDAS	Dynamic Data-Driven Applications Systems
DLT	Distributed Ledger Technology

DT	Digital Twins
FMaaS	Fair Mining as a Service
EPDS	Electric Propulsion Drive Systems
ICT	Information and Communication Technology
IoT	Internet of Things
LAN	Local Area Network
LO	Logical Object
MinT	Miner Twins
ML	Machine Learning
MoA	Microservice-Oriented Architecture
P2P	Peer-to-Peer
PBFT	Practical Byzantine Fault Tolerance
PO	Physical Object
PoB	Proof-of-Behavior
PoS	Proof-of-Stake
PoW	Proof-of-Work
SBC	Single Board Computer
SSA	Singular Spectrum Analysis
VR	Virtual Reality

References

1. Novo, O. Blockchain meets IoT: An architecture for scalable access management in IoT. *IEEE Internet Things J.* **2018**, *5*, 1184–1195. [CrossRef]
2. Nikouei, S.Y.; Xu, R.; Nagothu, D.; Chen, Y.; Aved, A.; Blasch, E. Real-time index authentication for event-oriented surveillance video query using blockchain. In Proceedings of the 2018 IEEE International Smart Cities Conference (ISC2), City, MO, USA, 16–19 September 2018; pp. 1–8.
3. Fitwi, A.; Chen, Y. Secure and Privacy-Preserving Stored Surveillance Video Sharing atop Permissioned Blockchain. *arXiv* **2021**, arXiv:2104.05617.
4. Xu, R.; Chen, Y. Fed-DDM: A Federated Ledgers based Framework for Hierarchical Decentralized Data Marketplaces. *arXiv* **2021**, arXiv:2104.05583.
5. Xu, R.; Zhai, Z.; Chen, Y.; Lum, J.K. BIT: A blockchain integrated time banking system for community exchange economy. In Proceedings of the 2020 IEEE International Smart Cities Conference (ISC2), Piscataway, NJ, USA, 28 September–1 October 2020; pp. 1–8.
6. Xu, R.; Chen, Y.; Blasch, E.; Chen, G. Exploration of blockchain-enabled decentralized capability-based access control strategy for space situation awareness. *Opt. Eng.* **2019**, *58*, 041609. [CrossRef]
7. Xu, R.; Chen, Y.; Blasch, E.; Chen, G. Blendcac: A blockchain-enabled decentralized capability-based access control for iots. In Proceedings of the 2018 IEEE International Conference on Internet of Things (iThings) and IEEE Green Computing and Communications (GreenCom) and IEEE Cyber, Physical and Social Computing (CPSCom) and IEEE Smart Data (SmartData), Halifax, NS, Canada, 30 July–3 August 2018; pp. 1027–1034.
8. Xu, R.; Chen, Y.; Blasch, E.; Chen, G. Blendcac: A smart contract enabled decentralized capability-based access control mechanism for the iot. *Computers* **2018**, *7*, 39. [CrossRef]
9. Barricelli, B.R.; Casiraghi, E.; Fogli, D. A survey on digital twin: Definitions, characteristics, applications, and design implications. *IEEE Access* **2019**, *7*, 167653–167671. [CrossRef]
10. Blasch, E.; Ravela, S.; Aved, A. *Handbook of Dynamic Data Driven Applications Systems*; Springer. 2018. Available online: https://link.springer.com/book/10.1007/978-3-319-95504-9#about (accessed on 15 November 2021).
11. Yaqoob, I.; Salah, K.; Uddin, M.; Jayaraman, R.; Omar, M.; Imran, M. Blockchain for digital twins: Recent advances and future research challenges. *IEEE Netw.* **2020**, *34*, 290–298. [CrossRef]
12. Xu, R.; Chen, Y.; Blasch, E. Microchain: A Light Hierarchical Consensus Protocol for IoT System. In *Blockchain Applications in IoT: Principles and Practices*; 2021. Available online: https://link.springer.com/chapter/10.1007/978-3-030-65691-1_9 (accessed on 15 November 2021).
13. Castro, M.; Liskov, B. Practical Byzantine fault tolerance. *OSDI* **1999**, *99*, 173–186.
14. Alsabah, H.; Capponi, A. Pitfalls of Bitcoin's Proof-of-Work: R&D Arms Race and Mining Centralization. SSRN 3273982. 2020. Available online: https://papers.ssrn.com/sol3/papers.cfm?abstract_id=3273982 (accessed on 15 November 2021). .
15. Nakamoto, S. *Bitcoin: A Peer-to-Peer Electronic Cash System*; Technical Report; Manubot. 2019. Available online: https://bitcoin.org/bitcoin.pdf (accessed on 15 November 2021).
16. Wang, W.; Hoang, D.T.; Hu, P.; Xiong, Z.; Niyato, D.; Wang, P.; Wen, Y.; Kim, D.I. A survey on consensus mechanisms and mining strategy management in blockchain networks. *IEEE Access* **2019**, *7*, 22328–22370. [CrossRef]
17. King, S.; Nadal, S. Ppcoin: Peer-to-peer crypto-currency with proof-of-stake. *Self-Publ. Pap. August* **2012**, *19*. Available online: https://decred.org/research/king2012.pdf (accessed on 15 November 2021).

18. Grieves, M. Digital twin: Manufacturing excellence through virtual factory replication. *White Pap.* **2014**, *1*, 1–7.
19. Erkoyuncu, J.A.; Butala, P.; Roy, R. Digital twins: Understanding the added value of integrated models for through-life engineering services. *Procedia Manuf.* **2018**, *16*, 139–146.
20. Van Schalkwyk, P. The Ultimate Guide to Digital Twins. 2019. Available online: https://xmpro.com/digital-twins-the-ultimate-guide// (accessed on 15 November 2021).
21. Khan, L.U.; Saad, W.; Niyato, D.; Han, Z.; Hong, C.S. Digital-Twin-Enabled 6G: Vision, Architectural Trends, and Future Directions. *arXiv* **2021**, arXiv:2102.12169.
22. Blasch, E.; Lambert, D.A. *High-Level Information Fusion Management and Systems Design*; Artech House. 2012. Available online: http://www.cs.utah.edu/~tch/notes/BRECCIA/refs/BOOK12_High-Level%20Information%20Fusion%20Management%20and%20Systems%20Design_BLASCH.pdf (accessed on 15 November 2021).
23. Bilberg, A.; Malik, A.A. Digital twin driven human–robot collaborative assembly. *CIRP Ann.* **2019**, *68*, 499–502. [CrossRef]
24. Sun, X.; Bao, J.; Li, J.; Zhang, Y.; Liu, S.; Zhou, B. A digital twin-driven approach for the assembly-commissioning of high precision products. *Robot. Comput.-Integr. Manuf.* **2020**, *61*, 101839. [CrossRef]
25. Karadeniz, A.M.; Arif, İ.; Kanak, A.; Ergün, S. Digital twin of egastronomic things: A case study for ice cream machines. In Proceedings of the 2019 IEEE International Symposium on Circuits and Systems (ISCAS), Sapporo, Japan, 26–29 May 2019; pp. 1–4.
26. Rassõlkin, A.; Vaimann, T.; Kallaste, A.; Kuts, V. Digital twin for propulsion drive of autonomous electric vehicle. In Proceedings of the 2019 IEEE 60th International Scientific Conference on Power and Electrical Engineering of Riga Technical University (RTUCON), Riga, Latvia, 7–9 October 2019; pp. 1–4.
27. Chen, X.; Kang, E.; Shiraishi, S.; Preciado, V.M.; Jiang, Z. Digital behavioral twins for safe connected cars. In Proceedings of the 21th ACM/IEEE International Conference on Model Driven Engineering Languages and Systems, Copenhagen, Denmark, 14–19 October 2018; pp. 144–153.
28. Kapteyn, M.G.; Knezevic, D.J.; Willcox, K. Toward predictive digital twins via component-based reduced-order models and interpretable machine learning. In Proceedings of the AIAA Scitech 2020 Forum, Orlando, FL, USA, 6–10 January 2020; p. 0418. Available online: https://arc.aiaa.org/doi/10.2514/6.2020-0418 (accessed on 15 November 2021).
29. Pargmann, H.; Euhausen, D.; Faber, R. Intelligent big data processing for wind farm monitoring and analysis based on cloud-technologies and digital twins: A quantitative approach. In Proceedings of the 2018 IEEE 3rd International Conference on Cloud Computing and Big Data Analysis (ICCCBDA), Chengdu, China, 20–22 April 2018; pp. 233–237.
30. El Saddik, A. Digital twins: The convergence of multimedia technologies. *IEEE Multimed.* **2018**, *25*, 87–92. [CrossRef]
31. Liu, Y.; Zhang, L.; Yang, Y.; Zhou, L.; Ren, L.; Wang, F.; Liu, R.; Pang, Z.; Deen, M.J. A novel cloud-based framework for the elderly healthcare services using digital twin. *IEEE Access* **2019**, *7*, 49088–49101. [CrossRef]
32. Laaki, H.; Miche, Y.; Tammi, K. Prototyping a digital twin for real time remote control over mobile networks: Application of remote surgery. *IEEE Access* **2019**, *7*, 20325–20336. [CrossRef]
33. Saad, W.; Bennis, M.; Chen, M. A vision of 6G wireless systems: Applications, trends, technologies, and open research problems. *IEEE Netw.* **2019**, *34*, 134–142. [CrossRef]
34. Nguyen, H.X.; Trestian, R.; To, D.; Tatipamula, M. Digital twin for 5G and beyond. *IEEE Commun. Mag.* **2021**, *59*, 10–15. [CrossRef]
35. Suhail, S.; Hussain, R.; Jurdak, R.; Oracevic, A.; Salah, K.; Hong, C.S. Blockchain-based Digital Twins: Research Trends, Issues, and Future Challenges. *arXiv* **2021**, arXiv:2103.11585.
36. Hassani, H. *Singular Spectrum Analysis: Methodology and Comparison*; 2007. Available online: https://mpra.ub.uni-muenchen.de/4991/ (accessed on 15 November 2021).
37. Dong, Q.; Yang, Z.; Chen, Y.; Li, X.; Zeng, K. Anomaly detection in cognitive radio networks exploiting singular spectrum analysis. In *International Conference on Mathematical Methods, Models, and Architectures for Computer Network Security*; Springer: New York, NY, USA 2017; pp. 247–259.
38. Xu, R.; Chen, Y.; Blasch, E.; Chen, G. A federated capability-based access control mechanism for internet of things (iots). In *Sensors and Systems for Space Applications XI*; International Society for Optics and Photonics: Bellingham, WA, USA, 2018; Volume 10641, p. 106410U.
39. Polunchenko, A.S.; Sokolov, G.; Du, W. Quickest change-point detection: A bird's eye view. *arXiv* **2013**, arXiv:1310.3285.
40. Yang, Z.; Chen, N.; Chen, Y.; Zhou, N. A novel PMU fog based early anomaly detection for an efficient wide area PMU network. In Proceedings of the 2018 IEEE 2nd International Conference on Fog and Edge Computing (ICFEC), Washington, DC, USA, 1–3 May 2018; pp. 1–10.
41. Moskvina, V.; Zhigljavsky, A. An algorithm based on singular spectrum analysis for change-point detection. *Commun. Stat.-Simul. Comput.* **2003**, *32*, 319–352. [CrossRef]
42. Hoecker, A.; Kartvelishvili, V. SVD approach to data unfolding. *Nucl. Instruments Methods Phys. Res. Sect. A Accel. Spectrometers Detect. Assoc. Equip.* **1996**, *372*, 469–481. [CrossRef]
43. Moskvina, V.; Zhigljavsky, A. Application of the Singular Spectrum Analysis for Change-Point Detection in Time Series. Ph.D. Thesis, Cardiff University, Cardiff, UK, 2001.

44. Nagothu, D.; Xu, R.; Chen, Y.; Blasch, E.; Aved, A. DeFake: Decentralized ENF-Consensus Based DeepFake Detection in Video Conferencing. In Proceedings of the IEEE 23rd International Workshop on Multimedia Signal Processing, Tampere, Finland, 6–8 October 2021.

45. Xu, R.; Nagothu, D.; Chen, Y. EconLedger: A Proof-of-ENF Consensus Based Lightweight Distributed Ledger for IoVT Networks. *Future Internet* **2021**, *13*, 248. [CrossRef]

46. Merkel, D. Docker: Lightweight linux containers for consistent development and deployment. *Linux J.* **2014**, *2014*, 2.

47. Golyandina, N.; Nekrutkin, V.; Zhigljavsky, A.A. *Analysis of Time Series Structure: SSA and Related Techniques*; CRC Press: Boca Raton, FL, USA, 2001.

48. Xu, R.; Nikouei, S.Y.; Nagothu, D.; Fitwi, A.; Chen, Y. BlendSPS: A BLockchain-ENabled Decentralized Smart Public Safety System. *Smart Cities* **2020**, *3*, 928–951. [CrossRef]

future internet

MDPI

Article

Utilizing Blockchain for IoT Privacy through Enhanced ECIES with Secure Hash Function

Yurika Pant Khanal [1], Abeer Alsadoon [1], Khurram Shahzad [1,*], Ahmad B. Al-Khalil [2], Penatiyana W. C. Prasad [1], Sabih Ur Rehman [1] and Rafiqul Islam [1]

[1] School of Computing, Mathematics and Engineering, Charles Sturt University, Melbourne 3062, Australia; yurikapant@gmail.com (Y.P.K.); alsadoon.abeer@gmail.com (A.A.); cwithana@csu.edu.au (P.W.C.P.); sarehman@csu.edu.au (S.U.R.); mislam@csu.edu.au (R.I.)

[2] College of Science, Department of Computer Science, The University of Duhok, Duhok 42001, Iraq; ahmad.al-khalil@uod.ac

* Correspondence: kshahzad@csu.edu.au

Abstract: Blockchain technology has been widely advocated for security and privacy in IoT systems. However, a major impediment to its successful implementation is the lack of privacy protection regarding user access policy while accessing personal data in the IoT system. This work aims to preserve the privacy of user access policy by protecting the confidentiality and authenticity of the transmitted message while obtaining the necessary consents for data access. We consider a Modified Elliptic Curve Integrated Encryption Scheme (ECIES) to improve the security strength of the transmitted message. A secure hash function is used in conjunction with a key derivation function to modify the encryption procedure, which enhances the efficiency of the encryption and decryption by generating multiple secure keys through one master key. The proposed solution eliminates user-dependent variables by including transaction generation and verification in the calculation of computation time, resulting in increased system reliability. In comparison to previously established work, the security of the transmitted message is improved through a reduction of more than 12% in the correlation coefficient between the constructed request transaction and encrypted transaction, coupled with a decrease of up to 7% in computation time.

Keywords: Internet of Things; blockchain; ECIES; secure hash function; privacy; reliability

Citation: Khanal, Y.P.; Alsadoon, A.; Shahzad, K.; Al-Khalil, A.B.; Prasad, P.W.C.; Rehman, S.U.; Islam, R. Utilizing Blockchain for IoT Privacy through Enhanced ECIES with Secure Hash Function. *Future Internet* **2022**, *14*, 77. https://doi.org/10.3390/fi14030077

Academic Editors: Rattikorn Hewett and Paolo Bellavista

Received: 10 January 2022
Accepted: 24 February 2022
Published: 28 February 2022

Publisher's Note: MDPI stays neutral with regard to jurisdictional claims in published maps and institutional affiliations.

1. Introduction

With the recent advances in technology, several Internet of Things (IoT) devices are being developed and implemented in our day to day life. These IoT devices collect personal data from the user to carry out different processes across several applications. Given the involvement of these devices in our daily life, the collected data are prone to a variety of security and privacy threats [1,2], in particular the monitoring of user's activities and profile creation [3]. Moreover, users do not have control over their data and necessary information regarding how it is being collected and how it is further processed. It thus becomes essential to protect the privacy rights of the users and facilitate them with the ability to control their transmitted data under the IoT landscape.

Data profiles can be utilised for individual identification purposes and therefore, collecting data and creating user data profiles pose a severe threat towards privacy and personal integrity. Even if the IoT data are not connected directly to an individual, it is possible to collect IoT data and create profiles of individuals. These profiles can be used to identify individuals or groups of individuals and pose a direct threat to user privacy. If data from IoT devices are combined with data from other sources such as social media, the identification of groups and/or individuals becomes much easier. One of the most critical parts of data collection via IoT devices is that most of the time, consumers are not aware of what data are being collected and how they are being used. Even in cases where consumers agree to the collection of data for a specific application, it is difficult for them to perceive the number of ways that data may be used in the future. The work of [4] investigates the possibilities to recognise a user based on when they communicate, what kind of applications they use, the type of devices they are surrounded by and their geographical location.

Traditionally, a user's sensitive data are stored on centralized servers [5], which can be easily tampered by the third party resulting in additional security and privacy threats, since user data was accessible without obtaining consent from the user. To address this issue, Blockchain-based solutions have been proposed in the IoT system, where several approaches have been advocated to protect user privacy [6–10]. Blockchain technology has dramatically enhanced user privacy and data access owing to its decentralized nature, enabling all participating nodes in the Blockchain to provide services equally [11]. In case of a node failure, other nodes keep providing the service, removing single point of failure that is a major problem in the traditional methods. The immutability feature of blockchain technology protects the data from being tampered and safely store the data in the form of blocks [12]. These features of blockchain technology eliminate the limitations of traditional centralized servers used in IoT applications. However, they still suffer from issues such as privacy protection and behavior regulation of access policy. In order to trace the real identity in an unusual transaction and preserve the privacy of the user in the data access policy, it is necessary to protect authenticity and confidentiality of the transmitted message while obtaining the consent needed for data access in the IoT system.

Our focus in this work is on protecting the confidentiality and authenticity of user consents during data transmission in IoT systems. We aim to preserve user privacy by maintaining the integrity of user consents before data transmission takes place in the IoT network. To improve the security strength of the encryption and decryption keys of the request transaction and response, we propose a two-pronged approach. Firstly, we proposed the use of a Secure Hash Function (SHF) [13] to derive private and public keys and secondly, we recommend the use of Key Derivation Function (KDF) to derive multiple keys to prevent the attacker from detecting the actual key value. The improved security strength decreases the correlation coefficient between constructed request transactions and encrypted transactions, enhancing user privacy in IoT systems. The proposed solution also improves the reliability of the system compared to a recent work of Lin et al. [14] by eliminating user-dependent variables and reducing the computation time.

The rest of the paper is organized as follows: Section 2 discusses in detail the recent advances in blockchain security measures, with a focus on its application in the IoT landscape. We detail the proposed scheme in Section 3, providing the major steps and associated details. Section 4 discusses the benefits of the proposed scheme, providing comparison to related works. In Section 5, we present analysis and detailed results of our scheme, demonstrating the efficacy in terms of average correlation coefficient and computation time, whereas the interim results on different datasets are also provided. Finally, the paper is concluded in Section 6, provisioning some future research directions.

2. Related Works

Blockchains are tamper evident and tamper resistant digital ledgers implemented in a distributed fashion, usually without a central authority. At their basic level, they enable a community of users to record transactions in a shared ledger such that under normal operation of the blockchain network, no transaction can be changed once published [15]. Unlike traditional methods, blockchain enables peer-to-peer transfer of digital assets without any intermediaries. Blockchain is often regarded as a public ledger in which all committed transactions are stored in a chain of blocks, and this chain continuously grows when new blocks are appended to it. The blockchain technology's key characteristics include decentralisation, persistency, anonymity and auditability.

2.1. Ethereum Public Blockchain

Ethereum represents a blockchain providing an abstract layer that enables all users to create their own rules for ownership, formats of transactions, and state transition functions, which is achieved through the use of smart contracts [16]. The consensus in the Ethereum network is based on modified GHOST protocol. Ethereum is created to tackle the issue of stale blocks in the network since the GHOST protocol includes stale blocks into calculations of the longest chain. The authors in [17] enhanced user privacy in a mobile crowdsensing system with spatial location privacy-preserving and greedy algorithms to improve data quality and preserve location privacy. They constructed a blockchain-based location privacy-preserving mobile crowdsensing system where the decentralization and immutability of Blockchain avoids security issues. However, the algorithm used in this scheme is based on the estimated value, so any inaccurate estimate may lead to significant problems and does not ensure data quality and reliability of the system.

In [18], the focus was on enhancing a smart healthcare system using a blockchain to preserve the privacy of the health data and ensure that diagnoses are not tempered. The proposed solution decreases the computation and communication cost comparing to the traditional system when preserving privacy in smart healthcare. However, the computation time is not fixed as the scheme requires users to update their key each time the transaction is updated. The researchers in [19] designed and implemented a decentralized reputation system to develop trust in the public fog nodes for enabling the IoT devices to rely on them securely. It provides safety against security vulnerabilities associated with IoT data and maintains the integrity of the data. The method uses the opinions of multiple users regarding the performance of public fog nodes to calculate reputation score for the future user to uses this system, which shows the unreliability of the system performance since the change in users' opinions changes the reputation score and increases the computation cost. Several computing task offloading schemes in mobile edge computing for IoT devices have been developed in [20]. The developed system uses a Blockchain-enabled edge computing framework and non-dominated sorting genetic algorithm to maintain data integrity while performing a task offloading process. Moreover, it adopts simple additive weighting and multi-criteria decision making techniques to select the most suitable offloading schemes. The task offloading system consumes 5% less energy than compared methods and decreases offloading time and energy consumption with data integrity and privacy protection. However, this work does not consider the security of VM instances while moving from one edge computing device to another device for obtaining load balance.

2.2. Consortium Blockchain

Consortium blockchain is a type of blockchain with authorized nodes to maintain distributed shared databases. Constructed by several organizations, the consortium blockchain is partially decentralized as only a small portion of nodes would be selected to determine the consensus. Among other advantages, recent works [21] have shown that it offers high potential for the establishment of decentralized electricity trading system with moderate cost. The authors of [22] propose a blockchain-based secure and privacy-preserving personal health information sharing scheme for diagnosis improvements in e-Health systems, where private and consortium blockchain are constructed by devising their data structures, and consensus mechanisms. In order to achieve data security, access control, privacy preservation and secure search in this work, all the data including the health information, keywords and the patients' identities are public key encrypted with keyword search. In [23], the authors construct a consortium blockchain framework for detecting malicious codes in malware and extracting the corresponding evidences in mobile devices. The work performs feature modelling by utilizing statistical analysis method, where the framework is composed of a detecting consortium chain shared by test members and a public chain shared by users. The authors also design a multi-feature detection method of

Android-based system for detecting and classifying malware, and establish a fact-base of distributed Android malicious codes by blockchain technology.

2.3. Hyperledger Fabric Blockchain

Hyperledger Fabric is an implementation of a distributed ledger platform for running smart contracts, leveraging familiar and proven technologies, with a modular architecture allowing pluggable implementations of various functions [24]. Designed as an extensible general-purpose permissioned blockchain, Hyperledger Fabric is the first blockchain system that supports the implementation of distributed applications written in standard programming languages [25]. This essentially allows them to be executed consistently across many nodes, giving impression of execution on a single globally-distributed blockchain computer, making Fabric the first distributed operating system for permissioned blockchains. The authors of [26] showed that the security can be enhanced by using proof of block and trade consensus algorithms to validate trade and blocks before allocating them to the ledger. Their solution uses a lightweight consensus algorithm, resulting in reduced computation time. However, it is resource intensive as it requires each trade to be validated before and at the time of block formation.

In [27], the authors proposed to improve privacy in industrial IoT with a Blockchain-based secure data sharing model for distributed multiple parties. They used federated learning algorithms to transform raw data generated in industrial IoT into the corresponding data model and share it. This model helps prevent data leakage, and data owners can assess before giving access to share their data in Industrial IoT. It provides high efficiency and enhanced security over traditional solutions. However, stable accuracy is difficult to achieve with the increase in the number of data providers. Also, an increase in the number of data providers requires a system to scale data for performing the computation. The consensus protocol is enhanced in [28] by checking the data loss before the data transmission to the blockchain network. This system uses a gossip-based diffusion function that guarantees the data collected from the sensor device are transmitted to the honest node of the blockchain network. However, this system does not consider the traffic that may increase in the network when the nodes are busy in replicating the processing outcome. The improvement of privacy with novel blockchain-based distributed key management scheme was discussed in [29], which eliminates the potential threat caused by a trusted third party. It uses multi-blockchain network that improves verification and saves storage space for IoT devices. The results showed that the scalability of the system is suitable to resource constrained IoT systems. However, a preshared key strategy in asymmetric cryptography is used, resulting in increased computation and communication overhead.

2.4. Blockchain Mechanisms for IoT Security

Blockchain-based frameworks to preserve user privacy in IoT have been proposed in a majority of works. The authors of [30] proposed a blockchain-based data acquisition scheme for a secure collection of data from IoT devices using Unmanned Aerial Vehicles (UAVs). This solution was researched by collecting data from IoT devices using UAV and storing safely in blockchain through mobile edge computing. However, in this approach, the required verification increases the latency. The researchers in [31] enhanced privacy in IoT with the Hyperledger Fabric Blockchain framework and Attribute Based Access Control (ABAC) to ensure efficient access control even under large number of requests in the IoT environment. The performance of this approach is analysed using two terminals which may increase the computational cost. The authors of [5] enhanced the publish/subscribe model with a blockchain-based secure publish/subscribe system to protect the privacy of publishers and subscribers. This model uses the Ethereum platform to ensure identity protection of the publisher and subscriber, using public key encryption with an equality test to guarantee the confidentiality of IoT data transmitted in the blockchain network. Though the authors present a promising way to preserve privacy in IoT system, the use of

Diffie–Hellman protocol for encryption procedure does not resist security attack, causing the user to compromise the security of their personal data.

Based on consortium blockchain, the security and privacy in IoT were enhanced in [32] with a novel attribute-based access control scheme. This scheme avoids the need to maintain an access control list in the IoT system as compared to traditional access control technologies. The access policies are made up of attributes and stored in the form of transaction in the blockchain. The performance analysis of their system shows storage overhead increases linearly with an increase in the number of attributes, whereas the computation overhead is also linear in the number of attributes. The security analysis shows that their scheme provides resistance to various security attacks in the IoT system. However, the key pair developed for authentication of the transaction does not boost the security strength of the encrypted transactions. In [14], the authors enhanced user privacy preservation in the IoT system with a novel secure mutual authentication system to provide traceability and privacy protection of access policy and user consent. The use of ECIES protects the confidentiality and privacy of request transaction message and response data that is transmitted to obtain necessary consents before data transmission in IoT. It gives a correlation coefficient of 0.34499 between constructed request transactions and encrypted transaction with a computation time of 102.733 ms. The ECIES is implemented to generate the public/private keys for encrypting and decrypting the request transaction data and response data. However, keys generated from the publicly exposed point on the elliptic curve result in violating user privacy.

A major concern regarding the adoption of blockchain technology in IoT networks is the enormous energy consumption associated with blockchains. This perception inevitably raises concerns about the further adoption of this technology, a fact that inhibits rapid uptake of what is widely considered to be a ground-breaking and disruptive innovation [33]. This fact, along with the significant increase in energy consumption caused by IoT networks has created a new challenge and diverted the focus towards creating an eco-friendlier IoT ecosystem, which provides energy efficient services and enables the production and use of renewable energy [34]. The combination of blockchains and a green IoT is focused on reducing energy consumption and adopting renewable resources rather than on energy generated by fossil fuels. Furthermore, recent studies [33,35] have shown that blanket statements about the energy consumption related to blockchains should be reviewed with care. Although Bitcoin and other proof-of-work blockchains do indeed consume a lot of power, alternative blockchain solutions with significantly lower power consumption are already available today, and new promising concepts are being tested that could further reduce the power consumption of large blockchain networks.

3. Modified ECIES with Secure Hash Function

The proposed scheme is intended to protect the integrity of transmitted messages while obtaining necessary consents for data transmission in IoT. Moreover, it provides resistance against different attacks and ensures reliable auditing of the user data access policy. To provide confidentiality and authentication of the transmitted data, both the request transaction and response data are authenticated once they are encrypted. We have chosen the proposed method in Lin et al. [14] as the basis for our designed solution. The mutual authentication system shows the access request transaction and response data while obtaining necessary consents. It protects against any data leakage and data loss, ensures reliable behavior auditing and protects the user access policy, preventing any malicious attack and possibility of consents versioning. The request transaction data are encrypted using ECIES and authenticated using message authentication code. The access request transaction and response data are firmly secured and authenticated, providing enhanced security while managing user data access policy and consents [18].

The use of an SHF to generate private and public keys prevents an attacker from detecting the actual values of the keys from which it is derived, even in the case where the hash function is known. This feature enhances the privacy preservation in IoT, providing resistance to detect the actual value of the key is used to encrypt the message. A detailed flow diagram of the proposed scheme is shown in Figure 1. In the following, we detail the major stages involved in our proposed scheme.

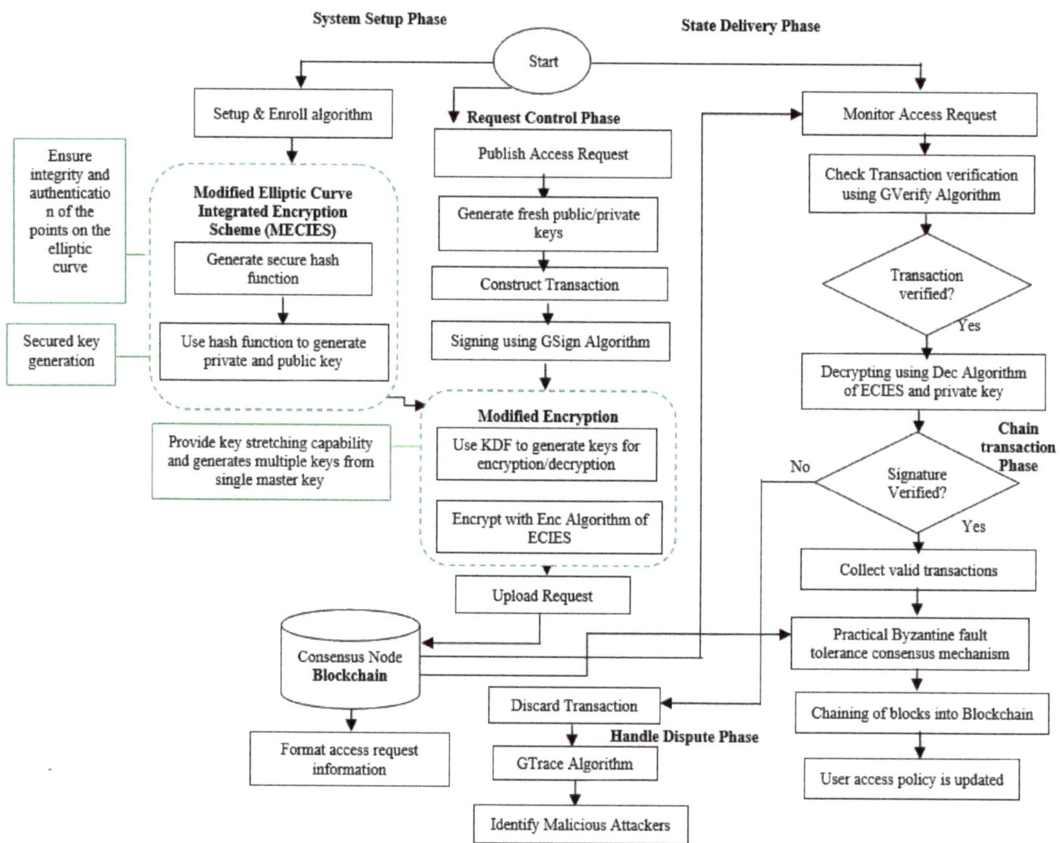

Figure 1. Flow Diagram of the Proposed Scheme—Modified ECIES with Secure Hash Function.

System Setup—The setup and enroll algorithm is invoked in this step to obtain keys for signing and verifying the transaction. After taking in the security parameters, λ^n, to obtain the public parameters, σ^n, we first generate a hash to ensure the security of the derived key, since the generated hash function is used to compute the private and public keys, denoted by δ_R and δ_P, respectively. The unique hash generation, corresponding to message m, is denoted as:

$$h(m) = \psi(m) \quad | \quad \psi : \{0,1\}^* \to \{0,1\}^{256},$$

where ψ is the unique hash generation function [30,36], and we have used SHA-256. In some works, for example, [14], the private and public key is calculated from publicly exposed points on the elliptic curve that can be easily detected by the attacker, and user privacy can be compromised. The security strength of the key ensures the confidentiality and authenticity of the transmitted message for obtaining the user consents before processing the user data.

The security strength of the key ensures the confidentiality and authenticity of the transmitted message for obtaining the user consents before processing user data in IoT.

Thus, if SHF is used to determine the value for the key rather than choosing publicly exposed points on the elliptic curve, the transmitted message will be highly protected. The secure hash value is used to generate the private key rather than randomly choosing a publicly exposed point on the elliptic curve as a private key and computing public key from the chosen private key. The private key δ_R is generated based on the hash function using the key generator function $\Gamma(\cdot)$, given as:

$$\delta_R = \Gamma(h) \quad | \quad \Gamma : \{0,1\}^* \rightarrow \{0,1\}^\kappa,$$

where κ is the designated key size [30,36]. After the generation of private key δ_R from the hash $h(m)$, corresponding to message m, the public key δ_P is calculated based on:

$$\delta_P = \delta_R * (E_x, E_y),$$

where (E_x, E_y) corresponds to the x and y coordinates of the point P on the elliptic curve E of finite field and P has the order of large prime number q [30]. Hence, the use of the hash function protects the value of the key being detected even if the hash function is known. As a result, private and public keys are secured and provide resistance to several security attacks enhancing user privacy protection.

Request Control—Once access request is published, new public and private keys are produced to avoid replay attack and profiling [14], where the uniquely generated hash is used to compute the private key instead of a randomly chosen key. The transaction to access the data is constructed and signed using the GSign algorithm. Request transaction data are then encrypted and verified using different keys. Since the randomly generated points on the elliptic curve can be detected by any attacker as multiple keys to encrypt the transmitted message, the proposed solution uses a KDF algorithm [37] to derive multiple keys from one secured master key. KDF follows an iterative process to derive multiple keys and ensure that an attacker is not able to identify origin of the master key [32]. After keys are generated, request transaction data are encrypted using the Enc. algorithm of ECIES and is authenticated using the MAC algorithm, where the encryption process is given by:

$$C_P = Encrypt(T_r, \delta_P),$$

and C_P represents the encrypted access request transaction data that is then uploaded to the blockchain network.

State Delivery—In this phase, consensus nodes in the blockchain network monitor the access request, checking the transaction verification using a signature verification algorithm. If the transaction is verified, it is decrypted using the Dec algorithm of ECIES and private key [14,30], which provides target device information and control orders, given by:

$$(D_i, C) = Decrypt(C_P, \delta_R),$$

and D_i and C represent the target device and control information, respectively. The consensus node of the blockchain network formats the data request to ensure the data access request is received from a valid requestor. The information access request to the user and response from the user is encrypted and authenticated. The authentication tag is recomputed to ensure the response is received from a valid user. If the authentication tag matches, only then the response from the user is decrypted to obtain response information about the request.

Chain Transaction—The transactions are retrieved in the smart contract of the blockchain network, where signatures are verified to check the validity of the transaction. If the transaction is valid, they are collected, and the block is formed. The consensus nodes use the Practical Byzantine Fault Tolerance consensus mechanism to chain the blocks [38]. The user access policy is then updated, which helps in managing consents set by the user.

Dispute Handling—The unusual transactions are traced by detecting abnormal and unusual behavior, where GTrace algorithm is executed to reveal the real identity in the unusual transactions. It helps to prevent impersonation attacks by identifying unusual behavior and showing the real identity of the attacker.

The flow of the modified ECIES is shown in Figure 2, whereas the steps of the proposed scheme are shown in Algorithm 1.

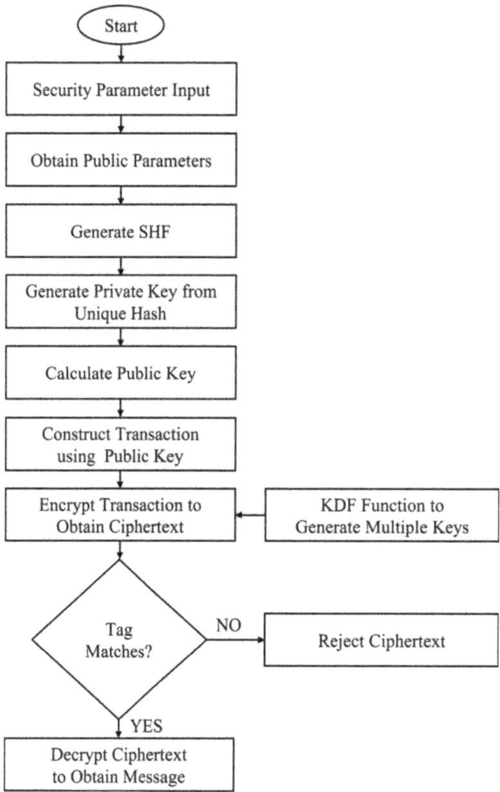

Figure 2. Flowchart of the Proposed Elliptic Curve Integrated Encryption Scheme with SHF.

Algorithm 1 Proposed ECIES with Secure Hash Utilization.

Input: Security parameter λ^n and Transactional Request Data T_r
Output: Response Data
1: **Generate** $\sigma^n \leftarrow \lambda^n$
2: **Compute** $h(m) \leftarrow m$
3: **Compute** $\delta_R \leftarrow h(m)$
4: **Compute** $\delta_P \leftarrow \delta_R * (E_x, E_y)$
5: **Construct** $T \leftarrow \delta_P$
6: **Encryption** $C_P \leftarrow Encrypt(T_r, \delta_P)$
7: **Authentication Check**
   if Tags Match
      $(D_i, C) \leftarrow Decrypt(C_P, \delta_R)$
   else Reject C_P
   end

Computation Time for Proposed Scheme

In this section, we calculate the computation time of the proposed scheme, which is given as:

$$T = T_b + T_c,$$

where T is the final computation time, T_b is a computation time for transaction generation and verification, and T_c is the initial computation time. Here, T_b is given as:

$$T_b = \sum_{i=1}^{T_r} T_r^i(t) + \sum_{i=1}^{\frac{N_s}{2}+1} N_s^i(t),$$

where $T_r^i(t)$ is a time for generation of one trade, and $N_s^i(t)$ is a time for verification by session node. Moreover, T_c is given by:

$$T_c = T_1 + T_h + T_2,$$

where T_h is the time for generation of hash function, and T_1 and T_2 correspond to the time of public/private key calculation and public parameter generation.

4. Benefits of Modified ECIES with SHF

The proposed solution helps improve the confidentiality and authenticity of the transferred message to obtain consents protected by using an SHF to generate private and public keys. This improves the correlation coefficient between transmitted messages and encrypted transactions. Along with this, it also ensures that the attacker is not able to detect the value of the key even in case hash function is known to the attacker because points on the elliptic curve are the order of a large prime number. In some of previous woks, the computation time is affected by the number of users, thus with the increase in the number of users, the computation time also increases, indicating the unreliability of the system. In the proposed scheme, the computation time is calculated by eliminating the user dependent variable, showing a higher system reliability.

SHF is utilized to generate private and public keys for improving the security strength of the transmitted message. The private key is generated from the SHF based on SHA-256, while the public key is calculated from the private key and points on the elliptic curve of the finite field that is the order of a large prime number. Hence, if the attacker tries to compute the point on the curve, they will not be able to detect the value of the key. In order to improve the efficiency of the encryption and decryption, the KDF is used to generate secured multiple keys from one master key. Some previous works [14] randomly select the publicly exposed point on the curve as a value of the key resulting in several security vulnerabilities that impact user privacy. Using publicly exposed points on the curve that are vulnerable to several attacks as a private and public key, will exploit the user privacy in IoT. Hence, the use of SHF will guarantee that the integrity of the key is protected, and the attacker is not able to detect the actual value of the key. In the proposed scheme, we have kept a regard for the authenticity and integrity protection of the transmitted message while consent management for enhancing user privacy in IoT by using ECIES with an SHF generation.

5. Results and Discussion

This section presents the analysis and results of the proposed scheme. Considering the relevance of Lin et al. [14] to our work, we provide a detailed comparison of our work with the results presented in Lin et al. [14]. MATLAB R2019a was used to implement and evaluate the prototype of the proposed model on a personal computer (PC). For the implementation, 'secp256r1' is used as the elliptic curve domain parameter [39] to develop the public parameter of the elliptic curve, whereas SHA-256 is used to secure the hash function generation. Four groups of 50, 150, 250, 500 device information were used as a dataset, where these datasets were taken from online resources [40]. Ten samples of device information from each group are taken to construct the request transactions. We considered attributes such as device_ID, device_Type, device_Model, and device_SN (serial number) from the device information for creating the tansaction request. The completed request transaction is encrypted and decrypted for both Lin et al. [14] and the proposed scheme. The strength of the transmitted message is measured in terms of the correlation coefficient between the constructed request transaction and encrypted request transaction. The performance evaluation of the proposed scheme is based on the comparison of correlation coefficient and computation time with that of Lin et al. [14].

We note that the correlation coefficient measures the closeness between the mapped points on the elliptic curve for the constructed request transaction and encrypted request transaction. The lower the value of the correlation coefficient, the more secure the encrypted transaction. We compared samples taken from our result with the device ID attribute of the 50-device group set from the dataset. This result consisted of the encrypted transaction for request transactions in the request control stage for both Lin et al. [14] and the proposed scheme, where the comparison is based on the correlation coefficient between constructed request transactions and encrypted request transactions.

Table 1 includes the device ID attribute of three samples; the constructed request transaction for each device ID and encrypted request transaction in Lin et al. [14] and our proposed scheme. The measured correlation coefficient here improves from 0.3451 to 0.3052 in the first sample of device ID attributes, which clearly demonstrates the improved security strength of the encrypted transaction due to the lower correlation coefficient. Apart from the device ID samples, we tested other attributes of the device information such as device_Type, device_Model, and device_SN attributes. Ten samples were taken from each of the datasets of 50-, 150-, 250-, and 500-device group set. The results are obtained during the request control stages before uploading the request transaction into the smart contract of the blockchain network, and are shown in Tables 2–5, respectively. It is evident from the provided tables that the proposed solution improves the correlation coefficient between the constructed request transaction and encrypted transaction, providing increased security strength of the encrypted transaction.

We also calculate the average values of the correlation coefficient and computation time for the proposed scheme and for Lin et al. [14], as shown in Table 6. The result shows a noticeable improvement in both the correlation coefficient and the computation time compared to Lin et al. [14]. Figure 3 shows the average correlation coefficient results for the proposed scheme and for Lin et al. [14], which demonstrates the security strength of the transmitted message. The results for Lin et al. [14] are shown in blue, while the orange color indicates the result for the proposed solution. Every paired blue-orange bar represents the correlation coefficient of the 50-, 150-, 250-, and 500- device group sets with the attributes device_ID, device_Type, device_Model, and device_SN, respectively. The average correlation coefficient for the proposed scheme for device_ID samples of the 50-device group dataset is reduced to 0.30122, whereas it is 0.34499 for Lin et al. [14]. Similarly, the average correlation coefficient for device_Model samples of 250-device group dataset is also reduced to 0.30359, whereas it is 0.34853 for Lin et al. [14]. Finally, the average correlation coefficient for device_SN samples of 500-device group dataset for the proposed solution is reduced to 0.30089 comparing to the record of 0.34433 for Lin et al. [14]. We attribute the degree of improvement in the correlation coefficient to the modified private and public keys for encryption in the proposed scheme. The proposed scheme improves the correlation coefficient from 0.04344 to 0.04377 between constructed request transactions and encrypted request transaction, which shows increased security strength of the transmitted message.

Table 1. Constructed Request Transaction and Encrypted Request Transaction in Lin et al. [14] and the Proposed Scheme.

Sample	Encrypted Request Transaction Samples from Lin et al. [14]		Encrypted Request Transaction Samples—Proposed Scheme	
Device_ID	Constructed Request Transaction	Encrypted Transaction	Constructed Request Transaction	Encrypted Transaction
5c504f2863	01∣∣pk1∣∣5c504f2863∣∣o	nMgxrrzzltep	01∣∣pk1∣∣5c504f2863∣∣o	#M25*^gh%@sEj_N
7j533g3785	01∣∣pk2∣∣7j533g3785∣∣r	VzBsirblemqxj	01∣∣pk2∣∣7j533g3785∣∣r	&2bgh?+5f*63^"bL+
2p488d4936	01∣∣pk3∣∣2p488d4936∣∣c	blskQohnerJk	01∣∣pk3∣∣2p488d4936∣∣c	Ox32?@><ghtSE21

Table 2. Correlation Coefficient and Computation Time Comparison of Lin et al. [14] and Proposed Scheme—Device ID Samples.

S. No.	Device_ID Samples	Constructed Request Transaction	Lin et al. [14]			Proposed Scheme		
			Encrypted Transaction	Correlation Coefficient	Computation Time (ms)	Encrypted Transaction	Correlation Coefficient	Computation Time (ms)
1	5c504f2863	01∣∣pk1∣∣5c504f2863∣∣o	nMgxrrzzltep	0.3451	108.45	#M25*^gh%@sEj_N	0.3052	97.87
2	7j533g3785	01∣∣pk2∣∣7j533g3785∣∣r	VzBsirblemqxj	0.3287	102.67	&2bgh?+5f*63^"bL+	0.2881	95.35
3	2p488d4936	01∣∣pk3∣∣2p488d4936∣∣c	blskQohnerJk	0.3695	110.88	Ox32?@><ghtSE21	0.3197	100.01
4	3r622h2678	01∣∣pk4∣∣3r622h2678∣∣w	kGniopHcqts	0.3586	105.5	&&4*^xo78?//@br	0.3074	97.36
5	8x923a0995	01∣∣pk5∣∣8x923a0995∣∣r	pxtrJvnerKlsgh	0.3218	100.3	Xx(+09%#<>P582j#	0.2821	90.8
6	5z307b2305	01∣∣pk6∣∣5z307b2305∣∣o	SzhioFnopsltr	0.3524	109.25	53>BJIO@+*29_ba	0.3117	99.3
7	1k408m7277	01∣∣pk7∣∣1k408m7277∣∣r	zcxvtDlfspqrv	0.3247	98.6	pM@0873##ghi++	0.2851	97.2
8	4v978x0355	01∣∣pk8∣∣4v978x0355∣∣r	QlnioghTsrvbe	0.3618	96.33	ST<**3789#(j;st_bt	0.3125	91.78
9	6g388k5669	01∣∣pk9∣∣6g388k5669∣∣o	twchjkioAans	0.3499	99.24	C5!(^78#"gmRb+523	0.3071	93.68
10	9s028n6082	01∣∣pk10∣∣9s028n6082∣∣c	ifniodfXtcrnig	0.3374	96.11	+93x0"^&pSq*?84((+	0.2933	91.45

Table 3. Correlation Coefficient and Computation Time Comparison of Lin et al. [14] and Proposed Scheme—Device Type Samples.

S. No.	Device_ Type Samples	Constructed Request Transaction	Lin et al. [14]			Proposed Scheme		
			Encrypted Transaction	Correlation Coefficient	Computation Time (ms)	Encrypted Transaction	Correlation Coefficient	Computation Time (ms)
1	Lamp	01 ǀ ǀpk1ǀ ǀlampǀ ǀo	hdlOxcjsmkbfaxb	0.3365	100.25	@2e78(^:xvyio#	0.2923	91.48
2	Fan	01 ǀ ǀpk2ǀ ǀfanǀ ǀc	IDvislzxkrFthjcs	0.3518	96.46	vM*{14s<"QJixh%j	0.3091	90.01
3	Air-conditioner	01 ǀ ǀpk3ǀ ǀacǀ ǀr	lpCivzodalfioeLt	0.3624	109.84	##hj89!kb(**vm%l	0.3147	101.21
4	Television	01 ǀ ǀpk4ǀ ǀtvǀ ǀr	glaQivtsjiwecbmf	0.3267	104.3	F4!{9(&&Hjck"b_1	0.2865	96.45
5	Freezer	01 ǀ ǀpk5ǀ ǀfreezerǀ ǀo	iozxJstovhgmcIDf	0.3378	98.8	Ox5%zkLR++8**d"	0.2934	93.26
6	Camera	01 ǀ ǀpk6ǀ ǀcameraǀ ǀc	bchjShBixmveloz	0.3413	97.65	++fg^*294(siX3!%K	0.2984	92.68
7	Doorbell	01 ǀ ǀpk7ǀ ǀdoorbellǀ ǀc	oxGjzbkdIvsohja	0.3649	95.38	&&59gX+jq6^^d!	0.3166	89.59
8	Door	01 ǀ ǀpk8ǀ ǀdoorǀ ǀr	mrXbjiwedjlHaMb	0.3672	94.71	3!cAm#]za!_vD8**	0.3193	89.45
9	Clock	01 ǀ ǀpk9ǀ ǀclockǀ ǀr	VbihKzrajioxbfk	0.3291	95.16	2!_xjdO(+"8fYios"	0.2891	89.78
10	Speaker	01 ǀ ǀpk10ǀ ǀspeakerǀ ǀo	aKleioshBzerjioc	0.3534	97.12	56@kWx"67!++^*8)	0.3112	90.56

Table 4. Correlation Coefficient and Computation Time Comparison of Lin et al. [14] and Proposed Solutions—Device Model Samples.

S. No.	Device_ Model Samples	Constructed Request Transaction	Lin et al. [14]			Proposed Scheme		
			Encrypted Transaction	Correlation Coefficient	Computation Time (ms)	Encrypted Transaction	Correlation Coefficient	Computation Time (ms)
1	RX350	01 ǀ ǀpk1ǀ ǀRX350ǀ ǀo	VbxdjklopStpd	0.3587	106.23	P#5!hbn2e<k"	0.3138	98.45
2	HS720A	01 ǀ ǀpk2ǀ ǀHS720Aǀ ǀc	rbpMiosgtkbdji	0.3393	109.04	++dfg*7D$%j{	0.2954	99.34
3	ZT8808	01 ǀ ǀpk3ǀ ǀZT8808ǀ ǀr	pbfKlacTrxkfdv	0.3718	113.96	J9_}ndb^&10f	0.3215	103.85
4	XY290P	01 ǀ ǀpk4ǀ ǀXY290Pǀ ǀr	Kgankobhmenx	0.3425	105.4	28g(7!kvy>?lb	0.2971	98.67
5	HDR6E	01 ǀ ǀpk5ǀ ǀHDR6Eǀ ǀo	AchjeoPvmftugy	0.3274	104.55	"fs9!45@kcql++	0.2887	96.77
6	CBT26Z	01 ǀ ǀpk6ǀ ǀCBT26Zǀ ǀc	ZxjdriobstJbci	0.3368	110.75	#46e%Jcmp8!(*	0.2932	101.48
7	PB485D	01 ǀ ǀpk7ǀ ǀPB485Dǀ ǀo	oxchksDLnfkwcy	0.3451	100.3	0x^{gno**57(%	0.2995	97.26
8	AVV56E	01 ǀ ǀpk8ǀ ǀAVV56Eǀ ǀr	GbjiochtgjFcodef	0.3596	104.78	rDk##99!hsi_4%!	0.3152	97.73
9	BM5060	01 ǀ ǀpk9ǀ ǀBM5060ǀ ǀc	abfdelUbjiotHny	0.3417	99.34	3!(gOx<@2dn+*>	0.2951	95.87
10	CR2030	01 ǀ ǀpk10ǀ ǀCR2030ǀ ǀo	rvpmRtzderighj	0.3624	102.45	+8cY{&269f##k!	0.3164	98.03

Table 5. Correlation Coefficient and Computation Time Comparison of Lin et al. [14] and Proposed Solutions—Device Serial Number Samples.

S. No.	Device_SN Samples	Constructed Request Transaction	Lin et al. [14]			Proposed Scheme		
			Encrypted Transaction	Correlation Coefficient	Computation Time (ms)	Encrypted Transaction	Correlation Coefficient	Computation Time (ms)
1	72020190805001	01ΙΙpk1ΙΙ72020190805001ΙΙr	cwkzAldOxvionc	0.3472	103.75	oxK*3#″4z89!Ws<k#	0.3072	97.33
2	72020190805002	01ΙΙpk2ΙΙ72020190805002ΙΙc	rcksiKlwgnoxhtVm	0.3381	107.22	@hs53!jL;(″bKx>++	0.2951	99.58
3	72020190805003	01ΙΙpk3ΙΙ72020190805003ΙΙr	MxjkdiyqosdGrdH	0.3564	109.55	##gP34{*oX629_jb*D	0.3115	102.67
4	72020190805004	01ΙΙpk4ΙΙ72020190805004ΙΙo	ldfivrskTaovhxGc	0.3415	105.14	″lB*{@793!_jf+>VG	0.2973	98.97
5	72020190805005	01ΙΙpk5ΙΙ72020190805005ΙΙo	bJoxjdlqieczgeorl	0.3261	99.34	PW(+*51U_″vz#A9<h	0.2861	96.88
6	72020190805006	01ΙΙpk6ΙΙ72020190805006ΙΙc	xjloFaicehpbhowc	0.3347	109.15	9^qxc*{_fk@bi56!	0.2937	101.45
7	72020190805007	01ΙΙpk7ΙΙ72020190805007ΙΙo	Lpwvnjxzaioerm	0.3641	99.62	++7Ox37″#bsT^y>*	0.3142	97.13
8	72020190805008	01ΙΙpk8ΙΙ72020190805008ΙΙr	mhykdgyerioskzt	0.3572	104.01	*fV%g_h!{″6do>&r	0.3116	98.35
9	72020190805009	01ΙΙpk9ΙΙ72020190805009ΙΙc	aQiocdjkguzpljXo	0.3487	98.15	Ix{&85+^dy@<>g#	0.3081	97.01
10	72020190805010	01ΙΙpk10ΙΙ72020190805010ΙΙr	PfsklchioxDgerzbj	0.3293	101.73	&jc*;31k4!+M_″*5%#	0.2841	96.78

Table 6. Average Correlation Coefficient and Average Computation Time Results of Lin et al. [14] and Proposed Scheme (from tested samples).

Dataset	Samples	No. of Tests Taken	Lin et al. [14]		Proposed Scheme	
			Average Correlation Coefficient	Average Computation Time (ms)	Average Correlation Coeffcient	Average Computation Time (ms)
50-Device Group Set	Device_ID	10	0.34499	102.733	0.30122	95.48
150-Device Group Set	Device_Type	10	0.34711	98.967	0.30306	92.447
250-Device Group Set	Device_Model	10	0.34853	105.68	0.30359	98.745
500-Device Group Set	Device_SN	10	0.34433	103.766	0.30089	98.615

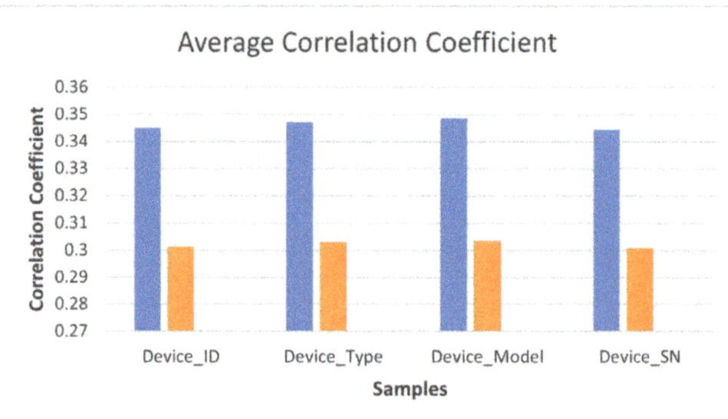

Figure 3. Average Correlation Coefficient results for Proposed Scheme and Lin et al. [14].

Figure 4 shows the average computation time results for both the proposed scheme and for Lin et al. [14] by calculating the execution time for each sample. The blue color indicates the results for Lin et al. [14], and the dark orange color indicates the result for the proposed solution. The paired blue-orange bars represent the average computation time for the 50-, 150-, 250-, and 500- device groupsets with the attributes device_ID, device_Type, device_Model, and device_SN, respectively.

- The average computation time for the proposed scheme of the device_ID samples of the 50-device group dataset is reduced to 95.48 ms, whereas it is 102.733 ms for Lin et al. [14];
- The average computation time for device_Type samples of 150-device group dataset is reduced to 92.447 ms compared to 98.967 ms of Lin et al. [14];
- The average computation time for device_Model samples of 250-device group dataset is 98.745 ms, which is less than the recorded value of 105.68 ms for Lin et al. [14];
- The average computation time for device_SN samples of 500-device group dataset. for the proposed solution is equal to 98.615 ms comparing to 103.766 ms for Lin et al. [14].

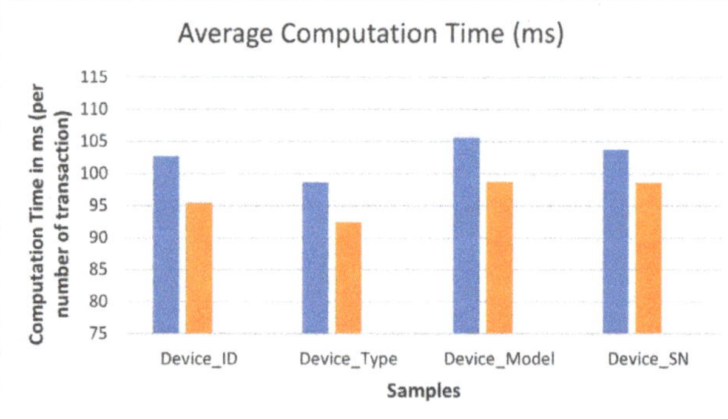

Figure 4. Average Computation Time results for Proposed Scheme and Lin et al. [14].

A comparison between our proposed scheme and Lin et al. [14] is presented in Table 7. Both solutions are based on ECIES that protect the confidentiality and privacy of request transaction messages and response data before data transmission in IoT. While Lin et al. [14] is mutually authenticated with ECIES, our proposed model modified the ECIES with an SHF. Using an SHF to derive private and public keys reduces the correlation coefficient, which improves the security strength of the request transaction data. Our contribution relies on the fact that SHF improves the strength of encryption/decryption of the transmitted message by adding new features for calculating private and public keys from the safer elliptic curve point, as compared to the case of Lin et al. [14], which does not use hash function generation in the process of calculating the private and public keys. Moreover, in Lin et al. [14], the security strength of the key was compromised, resulting in the violation of user privacy in IoT. However, to enhance the privacy and reliability of the processed user data in IoT, the new features adopted in the proposed scheme greatly enhance user privacy in the IoT system. The use of KDF during the encryption procedure of the request control stage introduces key stretching capability in the proposed scheme, which helps to derive multiple keys from a single master key. This feature decreases the number of iterations while deriving keys for authentication. As a result, the proposed scheme achieves a reduction in encryption and decryption time. The computation time calculated in the proposed scheme eliminates user dependent variables by including time for transaction generation and verification to calculate computation time. This feature ensures the reliability of the proposed scheme with reduced computation time compared to Lin et al. [14] by an average of 7 ms per number of transactions.

Table 7. Comparison between Proposed Scheme and Lin et al. [14].

Approach	Proposed Scheme Modified ECIES with a SHF	Approach of Lin et al. [14] Mutual Authentication with ECIES
Encryption/ Decryption Strength	The strength of the encryption/decryption is measured in terms of the correlation coefficient. The improvement in the correlation coefficient is from 0.34499 to 0.30122	Provides an average correlation coefficient of 0.34499.
Computation time	Computation time is measured in terms of execution time. The computation time decreases from 102.733 ms to 95.48 ms, reducing the encryption/decryption time from 39.925 ms and 41.513 ms to 34.444 ms and 35.859 ms.	Provide an average computation time of 102.733 ms with average encryption decryption time of 39.925 ms and 41.513 ms.
Contribution 1	The generation of an SHF increases the security strength of the key by adding new features for calculating private and public keys from the safer elliptic curve points. With the generation of an SHF, the security strength of the transmitted message is improved, which enhances the user privacy in IoT.	Does not use hash function generation for computing private and public keys for encrypting the transmitted message in IoT, which results in the violation of user privacy.
Contribution 2	The KDF introduces key stretching capability and decreases the number of iterations processes while deriving keys for authentication. This reduces the time for encryption and decryption.	The computation time is affected by the number of users showing the system unreliability.

6. Conclusions and Future Work

Data security and user privacy have been the emerging needs in the IoT system. In this work, we presented a Blockchain-based scheme to preserve user privacy in IoT. The proposed scheme provides a secure platform that allows the access requester to send the request transaction data and receive the response data for the corresponding request. We propose to use ECIES with SHF, which is the new feature adapted from Lin et al. [14], to protect the confidentiality and authenticity of the transmitted request transaction and response data. The use of an SHF to derive private and public keys enhanced user privacy in IoT. This enhancement could improve the security strength of the request transaction data, which helps to derive multiple keys from the single master key; decreasing the number of iterations while deriving keys for authentication and elimination. As a result, it reduces the computation time in the proposed solution by an average of 7ms per number of transactions compared to the work of Lin et al. [14]. In the future, we need to explore other cryptographic approaches to provide a secure platform for users and data requester to exchange their data in the IoT environment. Future research needs to focus on issues other than protecting the confidentiality and authenticity of the request transaction data and response data to enhance user privacy in IoT, such as investigating and utilizing different techniques to integrate within the blockchain network for achieving enhanced privacy in the IoT system.

Author Contributions: Conceptualization, Y.P.K.; Methodology, K.S.; Project administration, A.A. and S.U.R.; Resources, S.U.R.; Supervision, A.A., A.B.A.-K., P.W.C.P., S.U.R. and R.I.; Writing—original draft, Y.P.K. and K.S.; Writing—review & editing, K.S. All authors have read and agreed to the published version of the manuscript.

Funding: This research received no external funding.

Data Availability Statement: Not applicable, the study does not report any data.

Conflicts of Interest: The authors declare no conflict of interest.

References

1. Stoyanova, M.; Nikoloudakis, Y.; Panagiotakis, S.; Pallis, E.; Markakis, E.K. A survey on the internet of things (IoT) forensics: Challenges, approaches, and open issues. *IEEE Commun. Surv. Tuts* **2020**, *22*, 1191–1221. [CrossRef]
2. Sfar, A.R.; Natalizio, E.; Challal, Y.; Chtourou, Z. A roadmap for security challenges in the Internet of Things. *Digit. Commun. Netw.* **2018**, *4*, 118–137. [CrossRef]
3. Rantos, K.; Drosatos, G.; Kritsas, A.; Ilioudis, C.; Papanikolaou, A.; Filippidis, A.P. A blockchain-based platform for consent management of personal data processing in the IoT ecosystem. *Secur. Commun. Netw.* **2019**, *2019*, 1431578. [CrossRef]
4. Fernquist, J.; Fängström, T.; Kaati, L. IoT data profiles: The routines of your life reveals who you are. In Proceedings of the European Intelligence and Security Informatics Conference (EISIC), Athens, Greece, 11–13 September 2017; pp. 61–67.
5. Lv, P.; Wang, L.; Zhu, H.; Deng, W.; Gu, L. An IoT-oriented privacy-preserving publish/subscribe model over blockchains. *IEEE Access* **2019**, *7*, 41309–41314. [CrossRef]
6. Minoli, D.; Occhiogrosso, B. Blockchain mechanisms for IoT security. *Internet Things* **2018**, *1–2*, 1–13. [CrossRef]
7. Khan, M.A.; Salah, K. IoT security: Review, blockchain solutions, and open challenges. *Future Gener. Comput. Syst.* **2018**, *82*, 395–411. [CrossRef]
8. Alfandi, O.; Khanji, S.; Ahmad, L.; Khattak, A. A survey on boosting IoT security and privacy through blockchain. *Clust. Comput.* **2020**, *24*, 37–55. [CrossRef]
9. Roy, S.; Ashaduzzaman, M.; Hassan, M.; Chowdhury, A.R. Blockchain for IoT security and management: Current prospects, challenges and future directions. In Proceedings of the IEEE International Conference on Networking, Systems and Security (NSysS), Dhaka, Bangladesh, 18–20 December 2018; pp. 1–9.
10. Bisogni, C.; Iovane, G.; Landi, R.E.; Nappi, M. ECB2: A novel encryption scheme using face biometrics for signing blockchain transactions. *J. Inf. Secur. Appl.* **2021**, *59*, 102814. [CrossRef]
11. Hammi, M.T.; Hammi, B.; Bellot, P.; Serhrouchni, A. Bubbles of Trust: A decentralized blockchain-based authentication system for IoT. *Comput. Secur.* **2018**, *78*, 126–142. [CrossRef]
12. Gai, K.; Wu, Y.; Zhu, L.; Zhang, Z.; Qiu, M. Differential privacy-based blockchain for industrial internet-of-things. *IEEE Trans. Ind. Inform.* **2019**, *16*, 4156–4165. [CrossRef]
13. Gnatyuk, S.; Kinzeryavyy, V.; Kyrychenko, K.; Yubuzova, K.; Aleksander, M.; Odarchenko, R. Secure hash function constructing for future communication systems and networks. In Proceedings of the International Conference of Artificial Intelligence, Medical Engineering, Education, Moscow, Russia, 6–8 October 2018; pp. 561–569.
14. Lin, C.; He, D.; Kumar, N.; Huang, X.; Vijayakumar, P.; Choo, K.-K.R. Homechain: A blockchain-based secure mutual authentication system for smart homes. *IEEE Internet Things J.* **2019**, *7*, 818–829. [CrossRef]
15. Yaga, D.; Mell, P.; Roby, N.; Scarfone, K. Blockchain technology overview. *arXiv* **2019**, arXiv:1906.11078.
16. Buterin, V. Ethereum White Paper: A Next Generation Smart Contract & Decentralized Application Platform; 1st version. 2014; Volume 53. Available online: https://translatewhitepaper.com/wp-content/uploads/2021/04/EthereumOrijinal-ETH-English.pdf (accessed on 9 January 2022).
17. Zou, S.; Xi, J.; Wang, H.; Xu, G. Crowdblps: A blockchain-based location-privacy-preserving mobile crowdsensing system. *IEEE Trans. Ind. Inform.* **2019**, *16*, 4206–4218. [CrossRef]
18. Xu, J.; Xue, K.; Li, S.; Tian, H.; Hong, J.; Hong, P.; Yu, N. Healthchain: A blockchain-based privacy preserving scheme for large-scale health data. *IEEE Internet Things J.* **2019**, *6*, 8770–8781. [CrossRef]
19. Debe, M.; Salah, K.; Rehman, M.H.U.; Svetinovic, D. IoT public fog nodes reputation system: A decentralized solution using Ethereum blockchain. *IEEE Access* **2019**, *7*, 178082–178093. [CrossRef]
20. Xu, X.; Zhang, X.; Gao, H.; Xue, Y.; Qi, L.; Dou, W. BeCome: Blockchain-enabled computation offloading for IoT in mobile edge computing. *IEEE Trans. Ind. Inform.* **2019**, *16*, 4187–4195. [CrossRef]
21. Kang, J.; Yu, R.; Huang, X.; Maharjan, S.; Zhang, Y.; Hossain, E. Enabling localized peer-to-peer electricity trading among plug-in hybrid electric vehicles using consortium blockchains. *IEEE Trans. Ind. Inform.* **2017**, *13*, 3154–3164. [CrossRef]
22. Zhang, A.; Lin, X. Towards secure and privacy-preserving data sharing in e-health systems via consortium blockchain. *J. Med. Syst.* **2018**, *42*, 1–18. [CrossRef]
23. Gu, J.; Sun, B.; Du, X.; Wang, J.; Zhuang, Y.; Wang, Z. Consortium blockchain-based malware detection in mobile devices. *IEEE Access* **2018**, *6*, 12118–12128. [CrossRef]

24. Cachin, C. Architecture of the hyperledger blockchain fabric. In *Workshop on Distributed Cryptocurrencies and Consensus Ledgers*; 2016; Volume 310, pp. 1–4. Available online: https://allquantor.at/blockchainbib/pdf/cachin2016architecture.pdf (accessed on 9 January 2022).
25. Androulaki, E.; Barger, A.; Bortnikov, V.; Cachin, C.; Christidis, K.; Caro, A.D.; Enyeart, D.; Ferris, C.; Laventman, G.; Manevich, Y.; et al. Hyperledger fabric: A distributed operating system for permissioned blockchains. In Proceedings of the 13th EuroSys Conference, Porto, Portugal, 23–26 April 2018; pp. 1–15.
26. Biswas, S.; Sharif, K.; Li, F.; Maharjan, S.; Mohanty, S.P.; Wang, Y. PoBT: A lightweight consensus algorithm for scalable IoT business blockchain. *IEEE Internet Things J.* **2019**, *7*, 2343–2355. [CrossRef]
27. Lu, Y.; Huang, X.; Dai, Y.; Maharjan, S.; Zhang, Y. Blockchain and federated learning for privacy-preserved data sharing in industrial IoT. *IEEE Trans. Ind. Inform.* **2019**, *16*, 4177–4186. [CrossRef]
28. He, S.; Tang, Q.; Wu, C.Q.; Shen, X. Decentralizing IoT management systems using blockchain for censorship resistance. *IEEE Trans. Ind. Inform.* **2019**, *16*, 715–727. [CrossRef]
29. Ma, M.; Shi, G.; Li, F. Privacy-oriented blockchain-based distributed key management architecture for hierarchical access control in the IoT scenario. *IEEE Access* **2019**, *7*, 34045–34059. [CrossRef]
30. Islam, A.; Shin, S.Y. BUAV: A blockchain based secure UAV-assisted data acquisition scheme in Internet of Things. *J. Commun. Netw.* **2019**, *21*, 491–502. [CrossRef]
31. Liu, H.; Han, D.; Li, D. Fabric-IoT: A blockchain-based access control system in IoT. *IEEE Access* **2020**, *8*, 18207–18218. [CrossRef]
32. Ding, S.; Cao, J.; Li, C.; Fan, K.; Li, H. A novel attribute-based access control scheme using blockchain for IoT. *IEEE Access* **2019**, *7*, 38431–38441. [CrossRef]
33. Sedlmeir, J.; Buhl, H.U.; Fridgen, G.; Keller, R. The energy consumption of blockchain technology: Beyond myth. *Bus. Inf. Syst. Eng.* **2020**, *62*, 599–608. [CrossRef]
34. Sharma, P.K.; Kumar, N.; Park, J.H. Blockchain technology toward green IoT: Opportunities and challenges. *IEEE Netw.* **2020**, *34*, 263–269. [CrossRef]
35. Sedlmeir, J.; Buhl, H.U.; Fridgen, G.; Keller, R. Recent Developments in Blockchain Technology and their Impact on Energy Consumption. *arXiv* **2021**, arXiv:2102.07886
36. Hakeem, S.A.A.; Abd El-Gawad, M.A.; Kim, H. A decentralized lightweight authentication and privacy protocol for vehicular networks. *IEEE Access* **2019**, *7*, 119689–119705. [CrossRef]
37. Krawczyk, H. Cryptographic extraction and key derivation: The HKDF scheme. In *Annual Cryptology Conference*; Springer: Berlin/Heidelberg, Germany, 2010; pp. 631–648.
38. Sankar, L.S.; Sindhu, M.; Sethumadhavan, M. Survey of consensus protocols on blockchain applications. In Proceedings of the IEEE International Conference on Advanced Computing and Communication Systems, Coimbatore, India, 19–20 March 2017; pp. 1–5.
39. Brown, D.R. Sec 2: Recommended Elliptic Curve Domain Parameters. Standards for Efficient Cryptography, 2010. Available online: https://ci.nii.ac.jp/naid/10027922258/ (accessed on 9 January 2022).
40. Hang, L.; Kim, D.H. Design and implementation of an integrated iot blockchain platform for sensing data integrity. *Sensors* **2019**, *19*, 2228. [CrossRef] [PubMed]

future internet

MDPI

Article

Securing Environmental IoT Data Using Masked Authentication Messaging Protocol in a DAG-Based Blockchain: IOTA Tangle

Pranav Gangwani [1], Alexander Perez-Pons [1,*], Tushar Bhardwaj [2], Himanshu Upadhyay [2], Santosh Joshi [2] and Leonel Lagos [2]

[1] Department of Electrical & Computer Engineering, Florida International University, Miami, FL 33199, USA; pgang002@fiu.edu

[2] Applied Research Center, Florida International University, Miami, FL 33199, USA; tbhardwa@fiu.edu (T.B.); upadhyay@fiu.edu (H.U.); sajoshi@fiu.edu (S.J.); lagosl@fiu.edu (L.L.)

* Correspondence: aperezpo@fiu.edu

Abstract: The demand for the digital monitoring of environmental ecosystems is high and growing rapidly as a means of protecting the public and managing the environment. However, before data, algorithms, and models can be mobilized at scale, there are considerable concerns associated with privacy and security that can negatively affect the adoption of technology within this domain. In this paper, we propose the advancement of electronic environmental monitoring through the capability provided by the blockchain. The blockchain's use of a distributed ledger as its underlying infrastructure is an attractive approach to counter these privacy and security issues, although its performance and ability to manage sensor data must be assessed. We focus on a new distributed ledger technology for the IoT, called IOTA, that is based on a directed acyclic graph. IOTA overcomes the current limitations of the blockchain and offers a data communication protocol called masked authenticated messaging for secure data sharing among Internet of Things (IoT) devices. We show how the application layer employing the data communication protocol, MAM, can support the secure transmission, storage, and retrieval of encrypted environmental sensor data by using an immutable distributed ledger such as that shown in IOTA. Finally, we evaluate, compare, and analyze the performance of the MAM protocol against a non-protocol approach.

Keywords: IoT; security; privacy; environment; IOTA; Tangle; MAM; directed acyclic graph; blockchain; distributed ledger

Citation: Gangwani, P.; Perez-Pons, A.; Bhardwaj, T.; Upadhyay, H.; Joshi, S.; Lagos, L. Securing Environmental IoT Data Using Masked Authentication Messaging Protocol in a DAG-Based Blockchain: IOTA Tangle. *Future Internet* **2021**, *13*, 312. https://doi.org/10.3390/fi13120312

Academic Editor: Christoph Stach

Received: 22 October 2021
Accepted: 4 December 2021
Published: 6 December 2021

1. Introduction

1.1. Mobile and Electronic Environment

Current technological and economic advancements are exerting a tremendous influence on the environment, to the extent of raising severe concerns about climate change and pollution. Human activities have an undeniable and ever-increasing impact on the climate system, along with recent developments that are unprecedented and currently acknowledged by the Intergovernmental Panel on Climate Change [1]. Environmental monitoring in this context refers to an Internet of Things (IoT) system where sensors are used to collect useful data about the ecosystem, leading to further discoveries and a better and more comprehensive understanding, to execute specific actions in mitigating and addressing the degradation of the environment [2]. Environmental monitoring in indoor environments is another related field that is now gaining popularity. This has proved essential not only for the building's or housing's residents [3] but also in terms of lowering greenhouse gas emissions [4]. Temperature, humidity, rainfall, atmospheric pressure, light intensity, and air quality, which are impacted by pollutants such as carbon dioxide (CO_2), carbon monoxide (CO), sulfur oxide (SO_x), volatile organic compounds, and many more are among the most commonly measured parameters. CO is a gas that is colorless and odorless but can cause serious harm to the human population and to the environment. When a substance is burned, smoke and fumes are released containing CO. SO_x is a group of sulfur chemicals that cause serious harm to the environment. Therefore, monitoring of the environment and pollutants is required to achieve a safe and healthy environmental ecosystem [5].

To achieve the goal of a more distributed environmental ecosystem, algorithmic analysis of a large quantity of data [6] will train models that regularly monitor environmental parameters and notify or respond to anomalies in real-time. For the improvement of environmental monitoring, it is crucial to develop efficacious algorithms and models at a continued pace within the environmental community; fostering confidence in these methods requires publicly validated and verifiable processes. Within this system, devices can receive and submit data in a federated manner, with over-the-air updates as the shared algorithms, and models are enhanced with time.

The data used to train models must be tamper-proof and dependable, and the technique must be secure. This might help acquire the confidence of environmentalists and environmental officials, as well as make data collection for investigations easier. Environmentalists will have to look beyond their present methods to achieve this verified future. The rise of distributed ledger technology has the potential to close this gap. Therefore, we need to evaluate and analyze the performance of distributed ledger systems such as IOTA to securely exchange environmental data [7] and establish trust among environmental professionals.

1.2. Distributed Ledger Technologies

Distributed databases or distributed ledgers, such as blockchain [8], are managed via a consensus process by nodes in a peer-to-peer network. Despite the fact that all peers participate in maintaining database integrity, this consensus approach eliminates the necessity for a central administrator. Individuals may reclaim control over their data due to the lack of a central controller.

Blockchain was introduced in 2008 [9] as a distributed ledger technology that is decentralized and immutable. These attributes ensure that the data stored on the blockchain is secure, authentic, and distributed among all the peers in the network. There is no third party involved while making transactions on this technology and no central authority that can control it. These features open the door to various application domains and research areas such as IoT, healthcare, environment monitoring, AI, deep learning, security, and IoT data integrity, wherein the data needs to be distributed and tamper-resistant [10] to avoid a single point of failure when stored on a centralized database.

However, blockchain technology is facing several technical challenges despite having a great potential for the construction of future Internet systems [11]. The main challenge or concern is the scalability of blockchain. The time taken to mine a block is about 10 min and the block size is limited to 1 MB only. Moreover, the bitcoin blockchain is not able to deal with high-frequency trading since it is limited to 7 transactions per second. Additionally, the propagation of the blocks will be slow [12] if the block size is large, as it will require more storage space. Since few users would wish to maintain such a large blockchain, this will lead to centralization. Hence, it has been a tough challenge to address the tradeoff between block size and latency. Moreover, there is the possibility of selfish mining strategies, whereby miners can access greater rewards than they are entitled to.

In this paper, we have leveraged environmental sensory telemetry data, which consists of various sensory data parameters such as temperature, humidity, CO, liquid petroleum (LPG), smoke, light, and motion. The data was generated by a series of three customized sensor arrays. Sensor arrays, consisting of an MQ135 hazardous gas detection sensor, DHT22 temperature and humidity sensor, Onyehn IR pyroelectric infrared PIR motion sensor detector, and Anmbest light intensity detection photosensitive sensor were connected to Raspberry Pi sensory devices. Moreover, these devices were placed in distinct physical locations and variable environmental conditions, as shown in Table 1.

Table 1. Data description.

Column	Description	Units
Ts	Timestamp of event	Epoch
Device	Unique device name	String
CO	Carbon Monoxide	ppm (%)
Humidity	Humidity	Percentage
Light	Light detected?	Boolean
LPG	Liquid Petroleum Gas	ppm (%)
Motion	Motion detected?	Boolean
Smoke	Smoke	ppm (%)
Temp	Temperature	Fahrenheit

Each of these IoT devices is continuously collecting the sensory values from four sensors at a standard interval of 5 s. The data was collected during the span of "from 07/12/2020 00:00:00 UTC–07/19/2020 23:59:59 UTC" with a total number of "405,184 rows". In this framework, the "ISO standard Message Queuing Telemetry Transport (MQTT)" protocol [13] is leveraged to bind the sensory readings, a unique ID, and a timestamp and broadcast in terms of a single message; a sample payload is shown in Figure 1.

```
{
  "data": {
    "co": 0.006104480269226063,
    "humidity": 55.099998474121094,
    "light": true,
    "lpg": 0.008895956948783413,
    "motion": false,
    "smoke": 0.023978358312270912,
    "temp": 31.799999237060547
  },
  "device_id": "6e:81:c9: d4:9e:58",
  "ts": 1594419195.292461
}
```

Figure 1. Sample JSON payload.

Different consensus protocols and network topologies have been investigated; these are distributed to ensure the integrity of a distributed ledger while providing high transactions per second and zero fees for transactions. Algorand [14], IOTA [15], Hashgraph [16], and Ouroboros [17] are a few prominent protocols that promise to accomplish the aforementioned characteristics. This technology is suitable not just for the future of electronic finance but also for every data-driven industry.

In this paper, we propose an environmental monitoring [18] application of IOTA, which will allow environmental professionals, such as environmentalists and environmental officers, to share and store encrypted IoT sensor data in a secure way for monitoring purposes. IOTA is a permissionless distributed ledger protocol with no transaction fees. Its goal is to address the scalability concerns that have plagued previous distributed ledger technologies. Moreover, we have leveraged the "Masked Authenticated Messaging extension module of the IOTA protocol" in the proposed approach for the secure transmission, storage, and retrieval of encrypted environmental sensor data. The proposed approach is compared with another method without any data encryption protocol and the performance is measured in terms of time taken in the creation, attachment, and retrieval of payloads.

In summary, the contribution of this paper is as follows:

1. Design of and work using an environmental monitoring application that uses a DAG-based blockchain called IOTA to ensure the security and integrity of environmental IoT data.
2. Propose an architecture that uses IOTA nodes to implement an environmental monitoring application.
3. Implementation of a working model with its architectural design and an extensive evaluation of the model's performance with experiments.
4. Performance evaluation and comparison of the MAM protocol against a non-protocol approach with clear results.

The remainder of the paper is structured as follows. Section 2 illustrates the background technologies for the proposed framework. Section 3 contains the related literature review for choosing this technology and the proposed work. Section 4 highlights the detailed framework of the proposed model. Section 5 showcases the experimental setup, results, and discussion. Finally, Section 6 focuses on a conclusion and future directions in terms of optimizing the MAM protocol for securing sensory data.

2. Background

This section describes the various technologies utilized for the proposed work in this research.

2.1. IOTA

The IOTA is a distributed ledger technology to manage secure data transmission between different IoT devices. The main difference between the IOTA and other distributed ledger technologies is that it utilizes the directed acrylic graph (DAG) structure called the "Tangle" in place of the conventional blockchain. IOTA is highly scalable [19] since there are no blocks in its DAG structure, which leads to a faster confirmation of transactions, unlike in the case of blockchain. Making a transaction on IOTA consumes less energy [20] as compared to other distributed ledgers and, hence, the adoption of IOTA in low power devices such as the IoT becomes rudimentary.

The scalable architecture of the IOTA Tangle enables faster transaction confirmation, as shown in Figure 2. In Figure 2, each square represents a transaction and the arrows also known as edges connect these transactions to form a Tangle. There are three types of transactions, called tips, ongoing transactions, and approved transactions, as shown in Figure 2. Tips are the unconfirmed transactions that are new and have just been added to the ledger. Ongoing transactions are the transactions that have been added to the ledger and are waiting to be referenced by new transactions [21] to achieve confirmation. Approved transactions are the transactions that have been confirmed or have been referenced by all the tips, either directly or indirectly.

The working model and the security of the IOTA protocols were designed with quantum computers in mind, as well as environments with constraints on bandwidth. The Winternitz one-time signature system, which protects against quantum computer access, is used in the IOTA protocol. This one-time signature approach enables effective broadcast authentication in sensor networks since the communication and computing needs are low. As there is no transaction fee for publishing a transaction to IOTA, it can be seamlessly used to send transactions, store data, and ensure data integrity with time. A data transmission protocol [22] called masked authenticated messaging (MAM) enables a user to publish streams of encrypted data in the form of transactions. Participants can broadcast a message at any time by forming a channel [23]. Subscribers can subscribe to the channel of the publisher to receive the data by using the address of the transactions. However, a small amount of proof-of-work is necessary for the data to circulate through the network and prevent spamming. MAM allows the user to send encrypted data streams that are a chain of messages or sensor data to IOTA with zero cost per transaction through the Tangle.

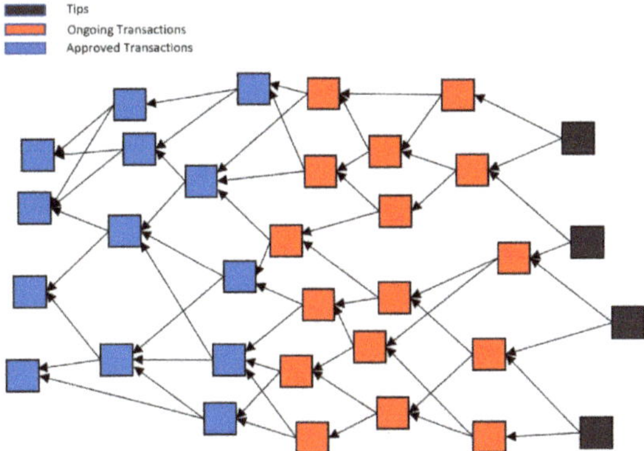

Figure 2. The DAG structure of the Tangle.

Forward secrecy and quantum-resistant cryptography are the two most important features of MAM implementation. Attacks by a quantum machine [24] that is adequately powerful can be resisted due to the secure post-quantum cryptographic algorithms. Many cryptographic algorithms traversing today over the Internet that are presently used to encrypt data [25] are not sufficiently secure. MAM is a useful protocol to transmit confidential data, due to the feature of forward secrecy. Every transaction is linked to the next transaction with a pointer known as next root, which is a Merkle root of the next transaction. As a result, the transaction at the point of entry and the subsequent transactions linked to it can be retrieved efficiently. However, it becomes infeasible for a user to fetch transactions before their point of entry due to forward transaction linking, as shown in Figure 3.

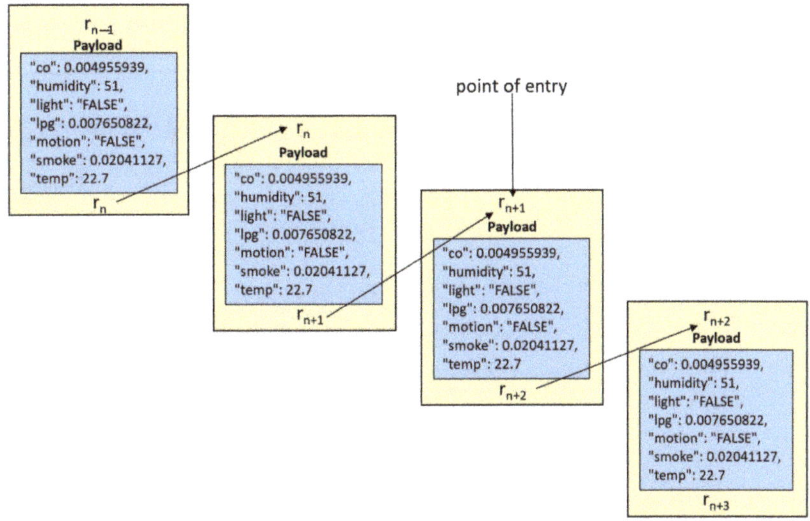

Figure 3. Transaction linking in MAM, displaying forward secrecy.

2.2. Modes and Channels of MAM

A channel is first established, then the publisher is able to encrypt data with the channel key and publish them into the Tangle. Clients can fetch the transaction from the Tangle and decode the message on it only if they know the MAM channel key. Messages are connected in chronological order and are published on the same channel. If the users gain access to a channel, they cannot view past transactions on that channel before their entry; this provides the notion of forward secrecy [26]. There are three modes of privacy provided by MAM, known as public, private, and restricted, that control visibility and access to the channels. The address of the transaction is the channel ID of MAM in each mode and allows the system to return a MAM transaction [27] by performing a straightforward request to the Tangle. In contrast, to decode the payload, the key provided in the transaction of MAM does not have to be the same as the channel ID. When the current payload is decoded, the user receives the message as well as the channel key for the subsequent message. For both private and public modes, this property becomes useful, as we will see below.

The channel key is the channel ID that makes up the transaction address for the public mode. Thus, all the contents of the message chain can be read by any user on the network. Due to the additional degree of protection, unauthorized users cannot read a message chain in private mode. The channel key is hashed [28], which becomes the channel ID as well as the transaction address. As a result, the channel key must be safely broadcasted to all subscribed users by the publisher in order that the message can be located on the Tangle network.

The next step involves the subscribed users obtaining the channel key's hash by querying the Tangle, using that key to decode the data payload. If an adversary intercepts a transaction of MAM sent in private mode, they will not be able to read or decode the content of the message payload by utilizing the channel ID, since it was produced by hashing the channel key.

Figures 4–6 represent the different channel modes of MAM and how transactions are linked. The root shown in the three figures is also known as the channel key; the address of the transaction is also known as the channel ID. In all three modes, as shown in the figures, each transaction contains a root and next root. The next root of the current transaction becomes the root for the next transaction, as shown. For the public mode, as shown in Figure 4, the transaction address or the channel ID is the same as that of the next root. For the private mode, as shown in Figure 5, the transaction address is the hash of the next root. However, in the restricted mode, as shown in Figure 6, there is an additional key, known as an authorization key, that is used for performing access control on the data. The transaction address for this mode is the hash of the next root, concatenated with the authorization key.

Public Mode

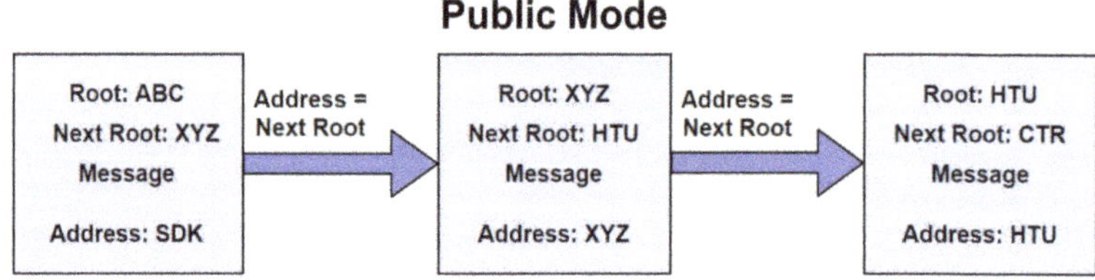

Figure 4. The flow of data in public mode.

Private Mode

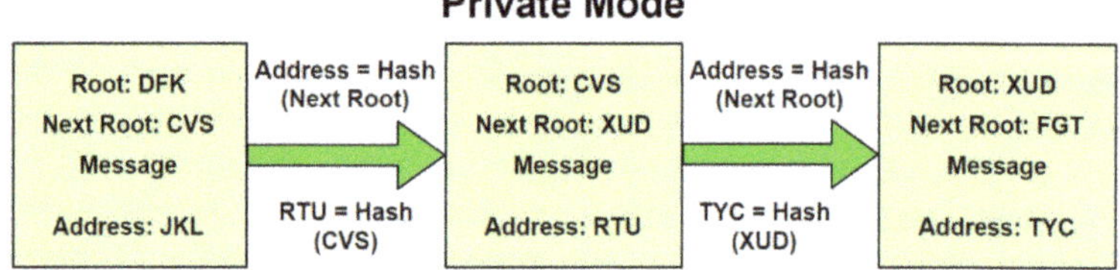

Figure 5. The flow of data in private mode.

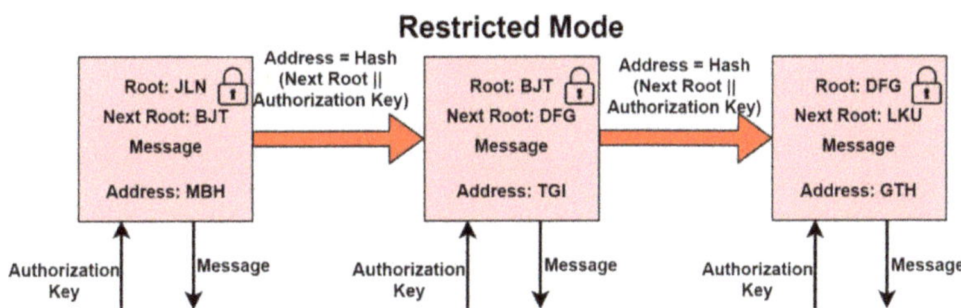

Figure 6. The flow of data in restricted mode.

Hashing of the authorization key is performed, and the hash is concatenated with the channel key to produce the transaction address of the restricted mode in MAM. The authorization key and the channel key are both necessary for decoding the data payload. The publisher specifies the authorization key and can change it at any time in the channel's stream. This enables the publisher to revoke access [29] in their channel from future messages at any point in time. If access is not granted to the subscriber for the current authorization key, they will be unable to decode and locate subsequent transactions in the chain of messages. As a result, subscribers' access can be revoked at any time using this approach. Figure 7 depicts a simplified representation of the many components that go into building an MAM channel.

Figure 7. Generation of channel key, using one-way hash functions.

3. Related Work

This section provides a detailed review of the literature in which blockchain technology and IOTA have been adopted in the domain of environmental monitoring and other IoT applications.

Bhandary et al. [30] present the use of a DAG-based blockchain structure called IOTA for the secure sharing of sensor data by integrating two technologies. The paper describes the work and the features of IOTA that can enable seamlessly integrating IoT devices with IOTA to safely transmit IoT data into the Tangle. An architecture was presented in the paper that included the use of Raspberry Pi devices to aggregate and send sensor data to the IOTA network. However, the architecture proposed in the paper was highly generic and lacked a working methodology. Moreover, there was no experimental evaluation of the architecture, especially in terms of performance.

Yu et al. [31] analyzed the stereotypical privacy and security issues in IoT and developed a framework that utilized Ethereum blockchain with an IoT system. A four-layered architecture was proposed where blockchain was used at the database layer to adapt to the IoT system. A good theoretical description of how the proposed framework tackles IoT security and privacy issues were provided. However, a proper working model to practically address these IoT security and privacy issues was missing. Furthermore, there was no performance or latency evaluation for the proposed framework.

Lamtzidis et al. [32] proposed a sensor node system that was distributed and utilized the IOTA distributed ledger to exchange data with IoT devices. In this paper, a distributed wireless sensor node system was proposed that ensured integrity of the data across the entire pipeline. The proposed system consisted of three entities: super nodes (SNs) that aggregated the data, full nodes (FNs), which are the IOTA nodes that perform the proof-of-work (PoW), and a back-end server. However, their proposed method did not show how the three entities are connected and how the data is flowing. Additionally, there was no implemented architecture and an experimental evaluation of their proposed system.

Yan et al. [33] proposed an environmental monitoring system that uses blockchain to provide integrity to the environmental data and prevent falsification. Additionally, a three-dimensional architecture using intelligent trusted devices was presented for environmental monitoring, to ensure the integrity and originality of the data collected by the IoT devices. The raw data from the sensors were transmitted to a whole node that could send data to the blockchain and could synchronize all the data within the blockchain nodes. However, an extensive performance evaluation of the proposed model was missing, which would provide some details on the latency of blockchain operations. Moreover, the traditional blockchain setup cannot meet the scalability demands of the IoT system; hence, a scalable blockchain or distributed ledger technology is required.

Shabandri et al. [34] presented an approach using the IOTA distributed ledger technology and IoT devices to demonstrate two IoT applications on the Tangle, such as a "smart utility meter system" and a "smart car transaction system". These proposed applications were connected to the internet using low power wide area networks (LPWAN). A DAG-based blockchain IOTA was used by the researchers to overcome the scalability and transactional cost of the conventional blockchain. Although the research paper gave detailed steps to implement the proposed applications, a well-defined architecture was missing and only a proof-of-concept (PoC) was presented.

Benedict et al. [35] proposed an implementation in the cloud that uses IoT-enabled blockchain to address some existing issues in smart cities. The research focuses on the use of "chaincodes", which are also known as smart contracts, for monitoring air quality systems in smart cities. An architecture called an "IoT-enabled blockchain for an air quality monitoring system (IB-AQMS)" was proposed and an experiment to assess the model was performed. However, the "chaincode" execution time for their approach was too high and would not satisfy the current IoT demands for a scalable system.

Guanochanga et al. [36] developed a wireless sensor network that monitored several air quality parameters within smart cities. An experiment was conducted on their proposed system and excellent results were obtained in the preliminary analysis. The preliminary results showed that the proposed approach could be used as a cost-effective tool for monitoring air quality. However, the approach lacked a framework or an entity that could ensure the integrity and security of the air quality data.

Mahmoud et al. [37] presented a review on the security of the IoT, various requirements for security, and proposed different countermeasures to secure IoT devices. A detailed description of the security issues that must be addressed at each layer of the IoT architecture was explained.

Bures et al. [38] provide a comprehensive review of the various features of IoT and the security challenges specifically related to IoT. The paper covered a vast number of security features and challenges that must be addressed to secure IoT devices and emphasized that security and privacy are the major security challenges that must be addressed to achieve a secure IoT system.

Our proposed architecture, which uses the IOTA nodes, overcomes the above-mentioned limitations. The proposed model satisfies the major security requirements for IoT, which include data confidentiality, integrity, and security at the application layer of the IoT stack or where the end-user requires the data. This provides a secure working environment for monitoring environmental IoT data generated from various IoT devices. Furthermore, in this paper, we conduct an extensive experimental evaluation of our proposed model to access its performance.

4. Proposed Architecture

The proposed work in this research paper aims to provide an environmental monitoring application by using the IOTA distributed ledger and the masked authenticated messaging (MAM) protocol. This application aims to ensure the security and privacy of the sensor data, as well as to control and prevent various environmental issues and hazards such as air pollution and greenhouse emissions. Moreover, this paper measures, analyzes, and compares the performance of the capability of MAM protocol using a non-protocol method.

4.1. Publishing and Fetching Environmental Sensor Data

We set out to evaluate MAM's potential for publishing environmental sensor data since it is a lightweight data communication protocol over an immutable distributed ledger. Using an MAM protocol, a system that could publish and fetch the environmental sensor data was developed. We installed the MAM Client JavaScript Wrapper library [39], as well as preparing the data payloads to be published to the private Tangle using MAM, and structured the data, utilizing the JSON format in the Windows client.

The Windows client was configured to publish the MAM data payloads through a restricted channel where a channel key and authorization key are used by the data publisher, i.e., an environmentalist, to encrypt the MAM data payloads. At the transaction level, an environmentalist can define the access controls. If an environmentalist wants to give one or more environmental officers access to their channels, they can send their channel keys to them. In return, the environmental officer could retrieve and authenticate the corresponding data payloads from the Tangle. If an environmentalist would like to revoke access to their stream of data at any time, this just requires updating their MAM channel's authorization key and safely transmitting it to a desired environmental officer.

With this architecture in place, as shown in Figure 8, the client device automatically published environmental sensor data to the private Tangle using the MAM Client JavaScript Wrapper library. Using MAM's restricted channel mode, data payloads were attached. We were able to examine how an environmentalist could change controls for accessing a specific stream of messages by upgrading their authorization key. To acquire the data payloads once the transactions were published to the Tangle, we used an authorization key and channel key.

We evaluated and compared the performance of our MAM implementation with a non-protocol-based approach, to further assess MAM's capability and applicability for this functionality. We published payloads, sized 145, 330, 515, and 740 KB, in the restricted channel configuration by utilizing the MAM client JavaScript Wrapper library for Node.js on the Intel(R) Core^{TM} i7-8565U processor of the Windows client device. We chose these sizes of payloads due to the limitations of the MAM protocol, which can handle a maximum size of 740 KB. Keeping the sensor data payloads, the processor, and the client device the same, we published the data payloads with a non-protocol approach [40] to the private Tangle, using Python and Jupyter Notebook as the runtime environment. Furthermore, we analyzed and compared the results of the two approaches.

4.2. Hashing of Merkle Tree

Hashed trees are generated using the Merkle hashing technique, where the direction of the trees goes upward. This tree is called the Merkle hash tree (MHT), wherein the leaves of the tree represent the hash of the values of the data or the ordered elements of a set. Let this authentic ordered set of elements for MHT be $x_{0,0}$, $x_{0,1}$, $x_{0,2}$, ..., $x_{0,n}$; therefore, the leaf node of the element $x_{0,i}$ will be the hash of that element. Let this leaf node be represented by $x_{1,i}$, where $x_{1,i} = H(x_{0,i})$ and $H()$ is a function that is cryptographically hashed one way.

A node in the MHT contains multiple incoming edges; the value of a node is the combined or concatenated hash [41] of its preceding nodes, also known as child nodes, where the sequence of the nodes is maintained. An internal node or a non-leaf node $x_{2,0}$ with child nodes $x_{1,0}$ and $x_{1,1}$ hence contains the value $x_{2,0} = H(x_{1,0}||x_{1,1})$. The MHT and a verification object that contains a set of nodes can be used to demonstrate the existence of an element. The root of the MHT [42] can be recomputed by the verifier by using the verification object and a set of nodes that are contained within it. The verifier compares the recomputed root using the verification object, with the publicly known root that the tree generates. For instance, consider the element $x_{0,0}$ in the MHT shown in Figure 9; the verification object consists of the values of the nodes $x_{0,0}, x_{1,1}$, and $x_{2,1}$. $x_{1,0} = H(x_{0,0})$, $x_{2,0} = H(x_{1,0}||x_{1,1})$ and conclusively, $root = H(y_{2,0}||x_{2,1})$ is constructed by the verifier. Once this verification object is constructed, the verifier can compare the computed root with the publicly known root and verify the value.

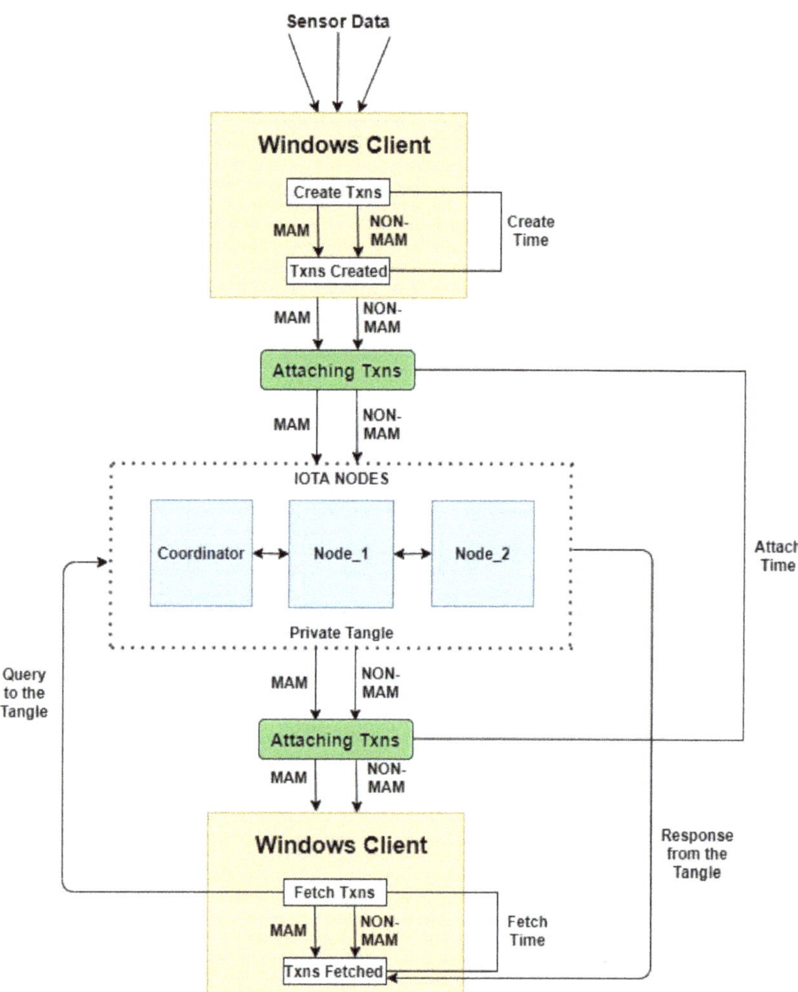

Figure 8. System architecture.

4.3. Hashing, Merkle Tree Signature Scheme, and One-Time Signatures

A digital signature technique is also known as the one-time signature (OTS) scheme can only be used to do a signature on one message with one key pair. Faster signing and algorithms for verification can be achieved with different techniques using hash-based OTS when compared to schemes such as RSA [43], which is a public-key digital signature technique. However, there are significant restrictions to OTS approaches, such as the length of signatures, the size of keys, and the maximum number of signatures possible.

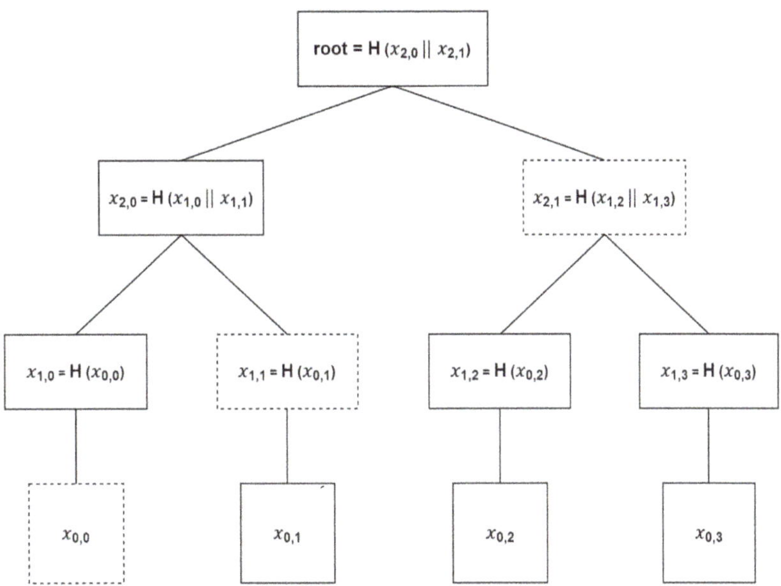

Figure 9. A binary Merkle hash tree built for the authentic values $x_{0,0}$, $x_{0,1}$, $x_{0,2}$, $x_{0,3}$. The values of nodes required to verify $x_{0,0}$ are bounded with broken lines.

Cryptographic secure hash functions ensure the security of OTS. The properties of a cryptographic secure hash function will be defined in this section. There are three categories, namely, "preimage-resistant", "second preimage-resistant" and "collision-resistant", and a hash function H: $\{0,1\}^*$ tends to $\{0,1\}^s$ that is cryptographically secure if it falls in the above three categories.

- Preimage-resistant.

For a hash function H, if it is hard to find any m for a given h with $h = H(m)$, then it is preimage-resistant.

- Second preimage-resistant.

For a hash function H, if it is hard to find any m_2 for a given m_1 with $H(m_1) = H(m_2)$, then it is second preimage-resistant.

- Collision-resistant.

For a hash function H, if it is hard to find a pair of m_1 and m_2 with $H(m_1) = H(m_2)$, then it is collision-resistant.

This multiple OTS can be verified by using a single public verification key, which is possible due to the MHT-based Merkle signature scheme (MSS). Each OTS scheme is represented by one leaf of the MHT. This implies that the same number of messages can be produced by each tree as the leaves of the MHT. The OTS technique's public verification keys [44] will be used to verify all of these communications. The OTS scheme's public verification keys are validated by computing the MHT's root from a specified verification object, as illustrated in Figure 7.

Only a limited number of messages can be signed with one public key, "pub_key", by using MSS. Let NUM = 2^n be the total possible number of messages since they must be a power of two. To generate the public key, "pub_key", the first step is to generate the private keys X_i and public keys Y_i of 2^n one-time signatures. For each public key Y_i, a hash value $H(Y_i)$ is calculated, where $1 \leq i \leq 2^n$. An MHT is constructed with these hash values,

h_i, $2^{n+1} - 1$ nodes, and 2^n leaves. In the MSS, the public key "pub_key" is the root of the Merkle tree.

Consider a message, M, which is to be signed with MSS [45]; first, a signature S results due to the signing, using a one-time signature technique on the message M. To execute the signature, S, one of the public and private key pairs (X_i, Y_i) is used [46]. Let the path from a given leaf to the root be denoted by P. The total number of nodes that path P contains is $n + 1$, with paths $P_1, \ldots, P_{n+1}$, where $P_1 = y_{1,i}$ is the leaf and $P_n = y_{n+1,0} = pub\_key$ is the root of the MHT. We require every child of the nodes $P_2, \ldots, P_{n+1}$ to compute P. As it is known that P_i is a child of P_{i+1}, therefore, to calculate the next node P_{i+1} of the path P, both the children of P_{i+1} must be known. To solve this computation, we require the sibling node of P_i. Let s_i be the sibling, such that $P_{i+1} = H(P_i||s_i)$ in the case where s_i is odd; if it is even, then $P_{i+1} = H(s_i||P_i)$. Therefore, n nodes, $s_0, \ldots, s_{n-1}$ are required to compute each node present in P. The signature of the MSS $sig = (S||s_2||s_3||s_{n-1})$ comprises the one-time signature S of the message M, plus the nodes.

The recipient knows the signature, $sig = (S||s_2||s_3||s_{n-1})$, the message M, and the public key pub_key. Firstly, the one-time signature S of message M is verified by the recipient. $P_1 = H(Y_i)$ is computed by the recipient, who hashes the public key of the one-time signature. For $k = 1, \ldots, n-1$, the nodes of P_k of path P are calculated with $P_k = H(y_{k-1}||s_{k-1})$ if the sibling index is odd, $P_k = H(s_{k-1}||y_{k-1})$ if it is even. The signature is valid if $P_n = pub\_key$ of the MSS.

5. Experimental Results and Analysis

An experiment was successfully performed that proved the feasibility of the proposed system to publish and retrieve authenticated, encrypted environmental IoT sensor data by using a distributed ledger. The MAM protocol ensured the source's validity and the data's integrity, which were formatted in the JSON format. We also demonstrated how an environmentalist might change the authentication keys to restrict permission to the data they may publish in the future. As a result, we demonstrated the potential of granular access controls, defined by the environmentalist.

A private Tangle was created, which consisted of three full nodes, called the coordinator, and two neighbor nodes, namely, "Neighbor Node_1" and "Neighbor Node_2" to test our proposed work, as shown in Figure 8. All three nodes were set up on three Linux Ubuntu servers with Hornet installed on them. Hornet is a powerful, community-driven IOTA node software written in the Go language and is a lightweight alternative to the IOTA reference implementation (IRI). Hornet was developed for the secure transfer of tokens or data, and for experimenting and implementing IOTA protocols between nodes or network participants. Machines can act as a node and connect to the IOTA network with the help of the Hornet software. These nodes or machines have functions such as authenticating the transactions, storing these authenticated transactions on the Tangle, and fetching these transactions back from the Tangle whenever required.

The dataset used was an open-source dataset that contained the environmental sensor data in a JSON format. The data included environmental parameters, such as temperature, timestamp, unique device id, carbon monoxide level, humidity percentage, light detected, liquified petroleum gas content, motion detected, and smoke levels. The payloads were created, and three actions (namely, create, attach, and fetch) were performed and analyzed in 300 trials. The Windows client machine, also known as the IOTA client, published and retrieved sensor data in the form of transactions to the Tangle. The client machine connects to the private Tangle using IOTA API and can make various API calls to perform various tasks. The experiment was performed with two approaches—(1) using the MAM protocol and (2) a non-protocol-based approach.

An experimental assessment was performed to evaluate the scalability of the two approaches. To achieve this, we focused on the three major tasks (i.e., create, attach, and fetch) that occur when publishing and fetching transactions. For the "create" task, we calculated the time it takes to create the transactions before publishing them to the IOTA nodes. The IOTA API was used to create the transaction object from the data payload and the execution time for this task was measured, which is called "create time".

The next step was to execute the "attach" task and calculate its execution time. Once the transaction object is created, it is published to the IOTA network by conducting the PoW and storing the transactions that the IOTA nodes perform. We calculated the execution time for this and labeled it as "attach time".

The final step was to execute the "fetch task" and calculate its execution time. After the transactions are published and stored in the IOTA network, we fetched these transactions by performing a query to the private Tangle, which in response provides the transactional data. We calculated the execution time for this fetch task and labeled it as "fetch time".

The three tasks can be mathematically expressed for the two approaches in the following way:

$$MAM_c = E(data) + ch\_gen$$
$$MAM_a = PoW(MAM_c) + stor(MAM_c)$$
$$MAM_f = D(que(MAM_a) + response)$$

$$NON\text{-}MAM_c = Enc(data)$$
$$NON\text{-}MAM_a = PoW(NON\text{-}MAM_c) + stor(NON\text{-}MAM_c)$$
$$NON\text{-}MAM_c = que(NON\text{-}MAM_c) + response$$

where E represents encryption; ch_gen represents channel generation; $stor$ represents storing; Enc represents encoding; and que represents a query.

5.1. MAM

The proposed work was performed using Node.js, an open-source cross-platform runtime environment for web application development [47]. Node.js apps are written in JavaScript and operate on a variety of platforms. MAM is an IOTA protocol that ensures only the authenticated parties are sending messages that are encrypted, ensuring both confidentiality and security. We used the restricted channel mode of MAM, which encrypts the data using the channel key and authorization key; only those parties having the correct keys can access the data from the IOTA Tangle.

With this system in place, the IOTA client created, attached, and fetched the sensor data payloads by executing the JavaScript code in Node.js, using the MAM Client JavaScript Wrapper library. After the transaction was published to the Tangle, the sensor data was fetched by using the channel key along with the authorization key. The results of this approach are displayed in Table 2.

Table 2. Results of the MAM experiment.

Processor	Action	Payload Size (KB)	Trials	Avg. Time (s)	St. Dev. (s)	Variance (s^2)	Min (s)	Max (s)
Intel(R) Core^{TM} i7-8565U	Create	145	300	0.49	0.13	0.017	0.415	1.34
Intel(R) Core^{TM} i7-8565U	Create	330	300	0.7	0.064	0.004	0.661	1.31
Intel(R) Core^{TM} i7-8565U	Create	515	300	0.95	0.026	0.001	0.91	1.07
Intel(R) Core^{TM} i7-8565U	Create	740	300	1.39	0.42	0.17	1.21	3.53
Intel(R) Core^{TM} i7-8565U	Attach	145	300	20.24	1.99	3.996	18.186	35.873
Intel(R) Core^{TM} i7-8565U	Attach	330	300	50.26	4.06	16.53	44.97	75.64
Intel(R) Core^{TM} i7-8565U	Attach	515	300	77.22	7.79	60.77	69.47	132.75
Intel(R) Core^{TM} i7-8565U	Attach	740	300	115.01	12.85	165.17	99.17	180.45
Intel(R) Core^{TM} i7-8565U	Fetch	145	300	0.426	0.032	0.001	0.361	0.697
Intel(R) Core^{TM} i7-8565U	Fetch	330	300	1.82	0.105	0.011	1.7	2.7
Intel(R) Core^{TM} i7-8565U	Fetch	515	300	1.92	0.138	0.019	1.79	2.94
Intel(R) Core^{TM} i7-8565U	Fetch	740	300	2.25	0.148	0.022	2.05	3.11

5.2. Non-Protocol Method

Using this approach, we published and retrieved the data payloads from the Tangle by only publishing the sensor data in the form of zero-value transactions [48], which are transactions that only contain data and no cryptocurrency, and without using any IOTA protocol. The proposed work was performed using Jupyter Notebook [49] which is an open-source web tool for creating and sharing documents with live code, equations, visualizations, and machine learning. Two Python scripts were written, where one script was configured to publish the sensor data, i.e., creating and attaching the transactions to the Tangle, while the other one was used to fetch the data from the Tangle. These two scripts utilized the official Python library for IOTA, called Pyota. Jupyter Notebook, running on the IOTA client, executed the two scripts to perform the abovementioned tasks.

With this system in place, the IOTA client published and fetched the sensor data from the Tangle with the help of the address whence the transactions were sent. The results of this approach are shown in Table 3.

We concentrated our investigation on the tasks that caused significant time delays for publishing and retrieving the messages. The three acts were studied: create, attach, and fetch. First, based on our findings, we discovered that the execution time for the message creation task was precise and was dependent on both the processor and the size of the payload. Second, we found that the average time for the attack process had a strong relationship with the size of the payload. Because the attaching stage involves the proof-of-work [50], which was conducted remotely by the private IOTA Tangle, a large variance and high correlation to payload size were expected. Thirdly, the average time to fetch a message from the private Tangle showed a high correlation to the payload size.

The average time was calculated for all three actions i.e., create, attach, and fetch, respectively, and are displayed in Figures 10–12. It can be seen that the MAM protocol performs far better than using any non-protocol method for publishing and retrieving sensor data from the private Tangle.

Table 3. Results of the non-protocol method.

Processor	Action	Payload Size (KB)	Trials	Avg. Time (s)	St. Dev. (s)	Variance (s^2)	Min (s)	Max (s)
Intel(R) Core^{TM} i7-8565U	Create	145	300	15.82	0.35	0.12	15.35	17.25
Intel(R) Core^{TM} i7-8565U	Create	330	300	29.03	0.62	0.38	27.88	31.49
Intel(R) Core^{TM} i7-8565U	Create	515	300	34.91	0.74	0.56	33.75	37.96
Intel(R) Core^{TM} i7-8565U	Create	740	300	91.75	1.73	2.99	88.76	97.07
Intel(R) Core^{TM} i7-8565U	Attach	145	300	23.3	0.11	0.012	22.97	23.59
Intel(R) Core^{TM} i7-8565U	Attach	330	300	52.91	0.11	0.013	52.45	53.2
Intel(R) Core^{TM} i7-8565U	Attach	515	300	82.53	0.18	0.033	81.87	82.95
Intel(R) Core^{TM} i7-8565U	Attach	740	300	118.3	0.68	0.471	116.4	119.6
Intel(R) Core^{TM} i7-8565U	Fetch	145	300	56.51	1.6	2.58	53.77	64.08
Intel(R) Core^{TM} i7-8565U	Fetch	330	300	135.1	38.4	6.2	159.2	127.2
Intel(R) Core^{TM} i7-8565U	Fetch	515	300	223.4	10	101	206.2	254.4
Intel(R) Core^{TM} i7-8565U	Fetch	740	300	327.9	5.99	35.9	318.1	346

5.3. Discussion

Since IoT devices are utilized in a variety of applications, there is a need to ensure data privacy and security, based on the application domain and the type of data being communicated between parties, such that an adversary cannot eavesdrop or tamper with the data.

Due to the IOTA distributed ledger, we achieved a tamper-proof audit trail of environmental sensor data, published from various IoT devices. The MAM extension module of IOTA provides environmentalists with the ability to publish, store and fetch the encrypted, authenticated, on-demand environmental sensor data by using the Tangle. The MAM protocol empowers the environmentalists by providing agency over the collected environmental sensor data, allowing them to share this data with the environmental officers for monitoring purposes. MAM's limited mode gives environmentalists fine-grained access controls over how data is shared across specialists in the digital environmental ecosystem, while the Tangle adds an extra layer of integrity to ensure that data is not tampered with. We discuss the privacy, security, and feasibility of our proposed system in the remainder of this section.

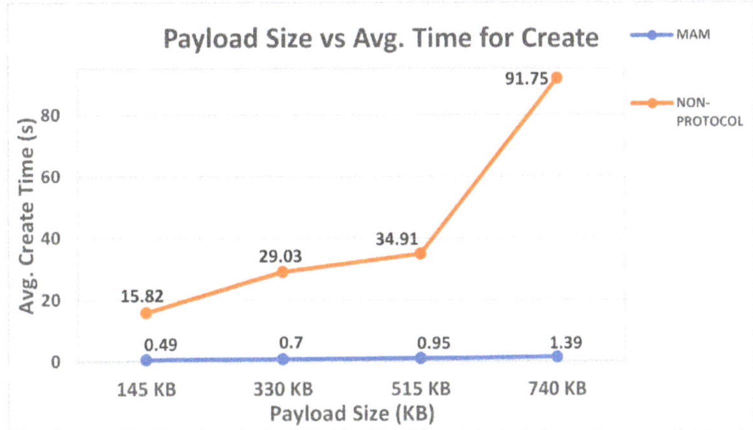

Figure 10. Line graph displaying the average "create time", with respect to payload size, for the MAM protocol and non-protocol methods.

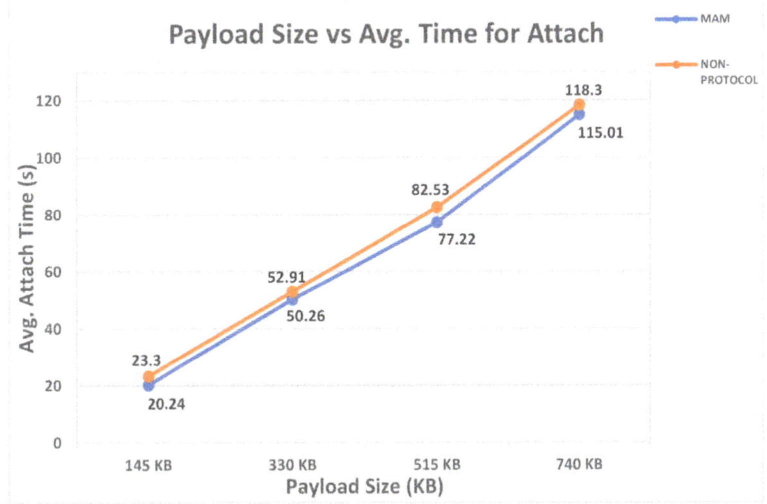

Figure 11. Line graph displaying the average "attach time", with respect to payload size, for the MAM protocol and non-protocol methods.

Since every node in a distributed ledger's network needs a copy of the current state of the ledger, distributed ledger technology seems to go against our present understanding of digital privacy. Despite the fact that value transactions on distributed ledgers may be pseudonymous, monitoring network traffic by analyzing the frequency of transactions and locations of origin could lead to the conclusion that one person has communicated with another regularly. Moreover, finding out the number of tokens an entity possesses is also possible, with varying levels of uncertainty. On a distributed ledger, enhancing privacy while maintaining auditability is still an ongoing area of research. Nevertheless, since MAM eliminates the concept of two entities communicating with each other, this issue does not pose a difficulty for our proposed system. Instead, the issuer generates transaction addresses at random in a data stream, regardless of who has access to the information required to decrypt the data. The following along of a public or private chain of messages can be achieved by the subscribers from their point of entrance forward since the next channel key is incorporated with the current message. Our technique, on the other hand, makes use of MAM's restricted mode, which, as mentioned in Section 2.2, allows an individual to revoke access from previous subscribers by making the addresses of future transactions unknown to them. This can be achieved only if the user changes the authorization key and, hence, access by undesired subscribers is revoked. Data can be kept private inside the transactions, due to the access controls that the MAM channel modes offer and are, thus, contradictory to the feature of transparency in distributed ledgers. If an environmentalist prefers not to have that amount of control over their data, they can generate and store authorization keys themselves, or they can delegate that power to environmental professionals with higher authority. The United States Environmental Protection Agency (US EPA) or another federal agency must store these authorization keys. Environmental officers will be allowed to access data if an environmentalist is unable to recollect them, give a log of their authorization keys, or, if the officer is needed to take immediate action, to manage an environmental threat.

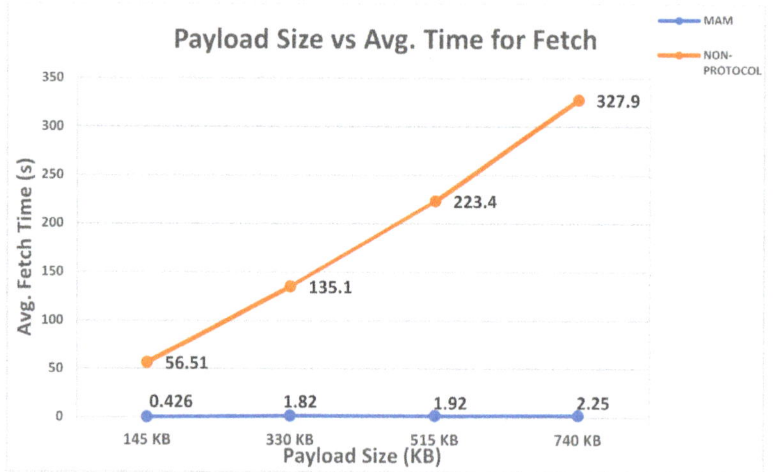

Figure 12. Line graph displaying the average "fetch time", with respect to payload size, for the MAM protocol and non-protocol methods.

Even though we demonstrated how MAM could be used to enable secure environmental sensor data exchange, we built our framework to be flexible enough to accommodate any open environmental data exchange standards. Furthermore, data can be transmitted using MAM from any endpoint with an internet connection, such as an environmentalist's computer, a server at a government institution like the US EPA, a mobile device, or a Bluetooth low-energy sensor. Our proposed system can be effortlessly linked with any professional in the digital environmental ecosystem due to the accessibility of encrypted data through open APIs. This, we believe, will facilitate acceptance, and open the door to new uses that go beyond environmental data collected without the oversight of environmental experts.

6. Conclusions and Future Directions

This study investigated the creation of an on-demand digital environmental ecosystem that relies on algorithms to analyze a huge amount of data, as well as the requirement that this data should be immutable, authenticated, and distributed. Using the MAM protocol, we demonstrated how encrypted environmental sensor data can be broadcasted, stored, and fetched from the IOTA Tangle to prove the data's integrity, security, and privacy. We also showed how granular access controls can be defined and updated by environmentalists. The MAM protocol proved to be a useful tool for encrypting and authenticating sensor data, although it may be improved in terms of performance and design. Many application fields, such as healthcare, the supply chain, and data storage of many kinds of sensor data, could benefit from this way of storing and delivering encrypted sensor data.

Based on the results of our extensive experimental evaluation of the proposed model, we can conclude that the MAM protocol performs better and provides better security than the non-protocol approach. The MAM protocol provided some additional features such as data encryption and granular access control, which provided better security and privacy, compared to the non-protocol method. Therefore, the MAM protocol can be seamlessly linked to various IoT devices to meet the scalability demands of these devices.

For future work, we suggest that the MAM protocol must integrate a secure and efficient key-transmitting method that would exchange the authorization keys between different entities. Additionally, as the MAM protocol develops and matures, we will demonstrate how this protocol can be used to ensure data integrity, by developing a proof-of-concept across academic universities.

Finally, we need to address how a huge dataset can be maintained across different stakeholders since the sensor data is widely distributed and the sensors are producing more data exponentially with time. IoT and embedded devices have the potential to generate massive amounts of data that will be incompatible with complete nodes, which cannot store the entire history of data. The complete nodes will keep track of the current state and prune the remaining data to make room for new transactions. An organization must have a complete record of all relevant transactions; nevertheless, the trimmed transactions will still have provable cryptographic links.

Author Contributions: P.G. created the proposed architecture and experimental scripts, and is the main writer. T.B. supervised the proposed model and is the secondary writer. S.J. contributed toward the writing of background technologies for this research. A.P.-P. and H.U. supervised the conducted research and reviewed the research article. Finally, L.L. provided the funding and the resources to perform this research. All authors have read and agreed to the published version of the manuscript.

Funding: This research was funded by the U.S Department of Energy National Energy Technology Laboratory (DOE NETL), grant number is DE-FE0031745.

Data Availability Statement: The dataset used for this research was publicly available and can be found here [https://www.kaggle.com/garystafford/environmental-sensor-data-132k (accessed on 20 July 2020)].

Conflicts of Interest: The authors declare no conflict of interest.

References

1. Mois, G.; Folea, S.; Sanislav, T. Analysis of Three IoT-Based Wireless Sensors for Environmental Monitoring. *IEEE Trans. Instrum. Meas.* **2017**, *66*, 2056–2064. [CrossRef]
2. Harris, M. Mules on a mountain. *IEEE Spectr.* **2016**, *53*, 50–56. [CrossRef]
3. Zhang, L.; Tian, F. Performance Study of Multilayer Perceptrons in a Low-Cost Electronic Nose. *IEEE Trans. Instrum. Meas.* **2014**, *63*, 1670–1679. [CrossRef]
4. du Plessis, R.; Kumar, A.; Hancke, G.; Silva, B. A wireless syste m for indoor air quality monitoring. In Proceedings of the IECON 2016—42nd Annual Conference of the IEEE Industrial Electronics Society, Florence, Italy, 23–26 October 2016; pp. 5409–5414.
5. Mukhopadhyay, S. Research activities on sensing, instrumentation, and measurement: New Zealand perspective. *IEEE Instrum. Meas. Mag.* **2016**, *19*, 32–38. [CrossRef]
6. Lee, M.; Offutt, A.J.; Alexander, R.T. Algorithmic analysis of the impacts of changes to object-oriented software. In Proceedings of the 34th International Conference on Technology of Object-Oriented Languages and Systems—TOOLS 34, Santa Barbara, CA, USA, 4 August 2000; pp. 61–70.
7. Lazarescu, M.T. Design of a WSN Platform for Long-Term Environmental Monitoring for IoT Applications. *IEEE J. Emerg. Sel. Top. Circuits Syst.* **2013**, *3*, 45–54. [CrossRef]
8. Dinh, T.T.A.; Liu, R.; Zhang, M.; Chen, G.; Ooi, B.C.; Wang, J. Untangling Blockchain: A Data Processing View of Blockchain Systems. *IEEE Trans. Knowl. Data Eng.* **2018**, *30*, 1366–1385. [CrossRef]
9. Nakamoto, S. Bitcoin: A Peer-to-Peer Electronic Cash System. *Decentralized Bus. Rev.* **2008**, 21260. [CrossRef]
10. Zheng, Z.; Xie, S.; Dai, H.N.; Chen, X.; Wang, H. Blockchain challenges and opportunities: A survey. *Int. J. Web Grid Serv.* **2018**, *14*, 352. [CrossRef]

11. Batubara, F.R.; Ubacht, J.; Janssen, M. Challenges of blockchain technology adoption for e-government. In Proceedings of the 19th Annual International Conference on Digital Government Research: Governance in the Data Age, Delf, The Netherlands, 30 May–1 June 2018; pp. 1–9.

12. Zachariadis, M.; Hileman, G.; Scott, S.V. Governance and control in distributed ledgers: Understanding the challenges facing blockchain technology in financial services. *Inf. Organ.* **2019**, *29*, 105–117. [CrossRef]

13. Yassein, M.B.; Shatnawi, M.Q.; Aljwarneh, S.; Al-Hatmi, R. Internet of Things: Survey and open issues of MQTT protocol. In Proceedings of the 2017 International Conference on Engineering & MIS (ICEMIS), Monastir, Tunisia, 8–10 May 2017; pp. 1–6.

14. Gilad, Y.; Hemo, R.; Micali, S.; Vlachos, G.; Zeldovich, N. Algorand. In Proceedings of the 26th Symposium on Operating Systems Principles, Shanghai, China, 28 October 2017; pp. 51–68.

15. Guo, F.; Xiao, X.; Hecker, A.; Dustdar, S. Characterizing IOTA Tangle with Empirical Data. In Proceedings of the GLOBECOM 2020–2020 IEEE Global Communications Conference, Taipei, Taiwan, 7–11 December 2020; pp. 1–6.

16. Akhtar, Z. From Blockchain to Hashgraph: Distributed Ledger Technologies in the Wild. In Proceedings of the 2019 International Conference on Electrical, Electronics and Computer Engineering (UPCON), Aligarh, India, 8–10 November 2019; pp. 1–6.

17. Kiayias, A.; Russell, A.; David, B.; Oliynykov, R. Ouroboros: A Provably Secure Proof-of-Stake Blockchain Protocol. In *Advances in Cryptology—CRYPTO 2017*; Springer International Publishing: Cham, Switzerland, 2017; pp. 357–388.

18. Othman, M.F.; Shazali, K. Wireless Sensor Network Applications: A Study in Environment Monitoring System. *Procedia Eng.* **2012**, *41*, 1204–1210. [CrossRef]

19. Silvano, W.F.; Marcelino, R. Iota Tangle: A cryptocurrency to communicate Internet-of-Things data. *Futur. Gener. Comput. Syst.* **2020**, *112*, 307–319. [CrossRef]

20. Silvano, W.F.; De Michele, D.; Trauth, D.; Marcelino, R. IoT sensors integrated with the distributed protocol IOTA/Tangle: Bosch XDK110 use case. In Proceedings of the 2020 X Brazilian Symposium on Computing Systems Engineering (SBESC), Florianopolis, Brazil, 24–27 November 2020; pp. 1–8.

21. Zivi, N.; Kadusic, E.; Kadusic, K. Directed Acyclic Graph as Tangle: An IoT Alternative to Blockchains. In Proceedings of the 2019 27th Telecommunications Forum (TELFOR), Belgrade, Serbia, 26–27 November 2019; pp. 1–3.

22. Bhandary, M.; Parmar, M.; Ambawade, D. Securing Logs of a System—An IoTA Tangle Use Case. In Proceedings of the 2020 International Conference on Electronics and Sustainable Communication Systems (ICESC), Coimbatore, India, 2–4 July 2020; pp. 697–702. [CrossRef]

23. Shafeeq, S.; Alam, M.; Khan, A. Privacy aware decentralized access control system. *Futur. Gener. Comput. Syst.* **2019**, *101*, 420–433. [CrossRef]

24. Ambainis, A.; Rosmanis, A.; Unruh, D. Quantum Attacks on Classical Proof Systems: The Hardness of Quantum Rewinding. In Proceedings of the 2014 IEEE 55th Annual Symposium on Foundations of Computer Science, Philadelphia, PA, USA, 18–21 October 2014; pp. 474–483.

25. Lee, H.K.; Malkin, T.; Nahum, E. Cryptographic strength of ssl/tls servers. In Proceedings of the 7th ACM SIGCOMM conference on Internet measurement—IMC '07, San Diego, CA, USA, 24–26 October 2007; p. 83.

26. Korotkyi, I.; Sachov, S. Hardware Accelerators for IOTA Cryptocurrency. In Proceedings of the 2019 IEEE 39th International Conference on Electronics and Nanotechnology (ELNANO), Kyiv, Ukraine, 16–18 April 2019; pp. 832–837.

27. Lamtzidis, O.; Pettas, D.; Gialelis, J. A Novel Combination of Distributed Ledger Technologies on Internet of Things: Use Case on Precision Agriculture. *Appl. Syst. Innov.* **2019**, *2*, 30. [CrossRef]

28. Ordieres-Meré, J.; Villalba-Díez, J.; Zheng, X. Challenges and Opportunities for Publishing IIoT Data in Manufacturing as a Service Business. *Procedia Manuf.* **2019**, *39*, 185–193. [CrossRef]

29. Nakanishi, R.; Zhang, Y.; Sasabe, M.; Kasahara, S. IOTA-Based Access Control Framework for the Internet of Things. In Proceedings of the 2020 2nd Conference on Blockchain Research & Applications for Innovative Networks and Services (BRAINS), Paris, France, 28–30 September 2020; pp. 87–95.

30. Bhandary, M.; Parmar, M.; Ambawade, D. A Blockchain Solution based on Directed Acyclic Graph for IoT Data Security using IoTA Tangle. In Proceedings of the 2020 5th International Conference on Communication and Electronics Systems (ICCES), Coimbatore, India, 10–12 June 2020; pp. 827–832.

31. Yu, Y.; Li, Y.; Tian, J.; Liu, J. Blockchain-Based Solutions to Security and Privacy Issues in the Internet of Things. *IEEE Wirel. Commun.* **2018**, *25*, 12–18. [CrossRef]

32. Lamtzidis, O.; Gialelis, J. An IOTA Based Distributed Sensor Node System. In Proceedings of the 2018 IEEE Globecom Workshops (GC Wkshps), Abu Dhabi, United Arab Emirates, 9–13 December 2018; pp. 1–6. [CrossRef]

33. Yan, J.; Zhang, F.; Ma, J.; An, X.; Li, Y.; Huang, Y. Environmental Monitoring System Based on Blockchain. In Proceedings of the 4th International Conference on Crowd Science and Engineering, Jinan, China, 18–21 October 2019; pp. 40–43.

34. Shabandri, B.; Maheshwari, P. Enhancing IoT Security and Privacy Using Distributed Ledgers with IOTA and the Tangle. In Proceedings of the 2019 6th International Conference on Signal Processing and Integrated Networks (SPIN), Noida, India, 7–8 March 2019; pp. 1069–1075. [CrossRef]

35. Benedict, S.; Rumaise, P.; Kaur, J. IoT Blockchain Solution for Air Quality Monitoring in SmartCities. In Proceedings of the 2019 IEEE International Conference on Advanced Networks and Telecommunications Systems (ANTS), Goa, India, 16–19 December 2019; pp. 1–6.

36. Guanochanga, B.; Cachipuendo, R.; Fuertes, W.; Benitez, D.S.; Toulkeridis, T.; Torres, J.; Villacis, C.; Tapia, F.; Meneses, F. Towards a real-time air pollution monitoring systems implemented using wireless sensor networks: Preliminary results. In Proceedings of the 2018 IEEE Colombian Conference on Communications and Computing (COLCOM), Medellin, Colombia, 16–18 May 2018. [CrossRef]
37. Mahmoud, R.; Yousuf, T.; Aloul, F.; Zualkernan, I. Internet of things (IoT) security: Current status, challenges and prospective measures. In Proceedings of the 2015 10th International Conference for Internet Technology and Secured Transactions (ICITST), London, UK, 14–16 December 2015; pp. 336–341.
38. Bures, M.; Klima, M.; Rechtberger, V.; Ahmed, B.S.; Hindy, H.; Bellekens, X. Review of Specific Features and Challenges in the Current Internet of Things Systems Impacting Their Security and Reliability. In *Trends and Applications in Information Systems and Technologies*; Springer International Publishing: Cham, Switzerland, 2021; pp. 546–556.
39. Zheng, X.; Sun, S.; Mukkamala, R.R.; Vatrapu, R.; Ordieres-Meré, J. Accelerating health data sharing: A solution based on the internet of things and distributed ledger technologies. *J. Med. Internet Res.* **2019**, *21*, e13583. [CrossRef] [PubMed]
40. Zhang, Y.; Nakanishi, R.; Sasabe, M.; Kasahara, S. Combining IOTA and Attribute-Based Encryption for Access Control in the Internet of Things. *Sensors* **2021**, *21*, 5053. [CrossRef] [PubMed]
41. Brogan, J.; Baskaran, I.; Ramachandran, N. Authenticating Health Activity Data Using Distributed Ledger Technologies. *Comput. Struct. Biotechnol. J.* **2018**, *16*, 257–266. [CrossRef] [PubMed]
42. Zhang, Y.; Wu, S.; Jin, B.; Du, J. A blockchain-based process provenance for cloud forensics. In Proceedings of the 2017 3rd IEEE International Conference on Computer and Communications (ICCC), Chengdu, China, 13–16 December 2017; pp. 2470–2473.
43. Zhou, X.; Tang, X. Research and implementation of RSA algorithm for encryption and decryption. In Proceedings of the 2011 6th International Forum on Strategic Technology, Harbin, Heilongjiang, 22–24 August 2011; pp. 1118–1121.
44. Mohassel, P. One-Time Signatures and Chameleon Hash Functions. In *Selected Areas in Cryptography*; Springer: Berlin/Heidelberg, Germany, 2011; pp. 302–319.
45. Buchmann, J.; Dahmen, E.; Klintsevich, E.; Okeya, K.; Vuillaume, C. Merkle Signatures with Virtually Unlimited Signature Capacity. In *Applied Cryptography and Network Security*; Springer: Berlin/Heidelberg, Germany, 2007; pp. 31–45.
46. Colavita, M.; Tanzer, G. A Cryptanalysis of IOTA's Curl Hash Function. 2018, pp. 1–13. Available online: https://www.boazbarak.org/cs127/Projects/iota.pdf (accessed on 3 December 2021).
47. Tilkov, S.; Vinoski, S. Node.js: Using JavaScript to Build High-Performance Network Programs. *IEEE Internet Comput.* **2010**, *14*, 80–83. [CrossRef]
48. Florea, B.C. Blockchain and Internet of Things data provider for smart applications. In Proceedings of the 2018 7th Mediterranean Conference on Embedded Computing (MECO), Budva, Montenegro, 10–14 June 2018; pp. 1–4.
49. Randles, B.M.; Pasquetto, I.V.; Golshan, M.S.; Borgman, C.L. Using the Jupyter Notebook as a Tool for Open Science: An Empirical Study. In Proceedings of the 2017 ACM/IEEE Joint Conference on Digital Libraries (JCDL), Toronto, ON, Canada, 19–23 June 2017; pp. 1–2.
50. Sarfraz, U.; Zeadally, S.; Alam, M. Outsourcing IOTA proof-of-work to volunteer public devices. *Secur. Priv.* **2020**, *3*, e98. [CrossRef]

future internet

MDPI

Review

CNN for User Activity Detection Using Encrypted In-App Mobile Data

Madushi H. Pathmaperuma, Yogachandran Rahulamathavan, Safak Dogan * and Ahmet Kondoz

Institute for Digital Technologies, Loughborough University London, London E20 3BS, UK;
o.m.h.pathmaperuma@lboro.ac.uk (M.H.P.); y.rahulamathavan@lboro.ac.uk (Y.R.); a.kondoz@lboro.ac.uk (A.K.)
* Correspondence: s.dogan@lboro.ac.uk

Abstract: In this study, a simple yet effective framework is proposed to characterize fine-grained in-app user activities performed on mobile applications using a convolutional neural network (CNN). The proposed framework uses a time window-based approach to split the activity's encrypted traffic flow into segments, so that in-app activities can be identified just by observing only a part of the activity-related encrypted traffic. In this study, matrices were constructed for each encrypted traffic flow segment. These matrices acted as input into the CNN model, allowing it to learn to differentiate previously trained (known) and previously untrained (unknown) in-app activities as well as the known in-app activity type. The proposed method extracts and selects salient features for encrypted traffic classification. This is the first-known approach proposing to filter unknown traffic with an average accuracy of 88%. Once the unknown traffic is filtered, the classification accuracy of our model would be 92%.

Keywords: encrypted traffic classification; network analysis; mobile data; network traffic to image

Citation: Pathmaperuma, M.H.;
Rahulamathavan, Y.; Dogan, S.;
Kondoz, A. CNN for User Activity
Detection Using Encrypted In-App
Mobile Data. *Future Internet* **2022**, *14*,
67. https://doi.org/10.3390/
fi14020067

Academic Editor: Christoph Stach

Received: 17 December 2021
Accepted: 15 February 2022
Published: 21 February 2022

Publisher's Note: MDPI stays neutral
with regard to jurisdictional claims in
published maps and institutional affiliations.

1. Introduction

In recent years, traffic classification has attracted increasing attention, as it is used in network management, security, advertising, network design, and engineering. Network traffic classification involves analyzing traffic flows and identifying the type of content within these flows. In network traffic analysis, a network trace of a device or a group of devices is taken as input and, as output, information about those devices, their users, their apps, or in-app activities is given. Network traffic classification has many possibilities to solve personal, business, internet service provider, and government network problems such as anomaly detection, quality of service control, application performance, capacity planning, traffic engineering, trend analysis, interception, and intrusion detection. To date, several traffic classifications approaches have been proposed and developed. These methods have evolved significantly over time from port-based, deep packet inspection (DPI) to machine learning (ML) methods [1]. The use of dynamic port-negotiation mechanisms by applications has made port-based methods no longer suitable. An increase in the use of encrypted internet traffic and privacy policies that prevent access to packet content have rendered relying on a packet's payload or DPI no longer effective. Most research activities that perform encrypted traffic classification rely on extracting statistical features from traffic flows, which is followed by performing feature selection to eliminate irrelevant and redundant features. They then use classical machine learning algorithms, such as random forest (RF) [2,3], Bayes net [4], K-nearest neighbors (KNNs), and support vector machine (SVM) [5] to perform the classification. These methods can handle both encrypted and unencrypted traffic. However, the performance of these methods greatly relies on human-engineered features, which limit their generalizability [1].

CNNs, an important model of deep learning (DL), were initially applied in the field of image recognition [6] and achieved remarkable results. With the advances in DL, the use of CNNs has become prevalent in many fields such as speech recognition, audio processing, visual document analysis, genomics, and in medical use [7]. Yet, it has not been sufficiently utilized in network traffic classification [6]. This paper proposes an image-based method that represents network traffic as images and utilizes DL architecture based on CNNs to learn the traffic features in these images and perform traffic classification. This research focused on conducting network traffic classification to identify user activities performed on mobile applications (known as in-app activities) from a sniffed encrypted internet traffic stream. Figure 1 shows the generated images using encrypted in-app activity data for three different in-app activities. The main advantage of using a CNN in image classification is that there is no need to extract features beforehand, because the CNN model can learn features by itself.

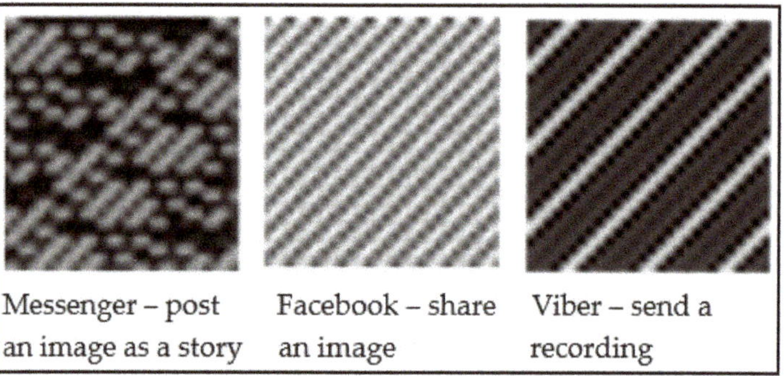

Figure 1. Grayscale images generated from the encrypted in-app activity data.

In this research, sensitive information related to online users, such as the activities performed with their mobile apps, are inferred by passively sniffing encrypted wireless network traffic. Even though encryption protocols are used to encrypt data, it protects only the packet's payload, but it does not hide *side channel data* such as frame length, data length, inter arrival time, and direction (incoming/outgoing). Therefore, in this research, the side channel data were used to reveal private information related to the user's online behavior. Users perform different activities using the apps installed on their mobile devices. Each in-app activity has a distinct network behavior [8] and, thus, generates different traffic flows. Traffic flow data are converted into images and image classification deep learning techniques are used to detect in-app activities.

The contributions of this paper are summarized as follows:

(a) Unknown app data detection: A wide range of apps are available in app stores. It is not practical to train a machine learning model for such a wide variety of apps and for every in-app activity that can be performed using those apps. When deploying an in-app activity detection model in a real network, the model needs to correctly detect applications and in-app activities while earmarking previously untrained activities as unknown traffic. Most existing works in the literature involve model training and testing on the same set of apps, which renders them unfit for detecting previously unknown traffic. The proposed model in our work was designed to conduct network traffic analyses effectively, even in the presence of noise generated by unknown traffic;

(b) Activity detection using minimal data: When network traffic is captured to detect in-app activities, there can be instances when an eavesdropper captures only part of the activity-related traffic instead of the entire transaction, as the target user's activity may have been underway already. The proposed model was designed to identify in-app activities even by observing a subset of an activity's traffic;

(c) Fine-grain in-app activity detection: Works in the literature have considered coarse-grained activities such as uploading, chatting, downloading, or generic activities, including sending an email, messaging on Messenger, and posting on Instagram. Our work advances this existing body of work by designing the model to identify fine-grained in-app activities. For example, when a generic Facebook activity "posting on wall" is considered, the model can determine whether the post is an image, a long text, a short text, a video, or a check-in. This level of fine-grained analysis is challenging, as classification is performed in an encrypted domain using only side channel data;

(d)   Novel data set for future research: A comprehensive data set was created by performing a series of actions on eight apps, namely, Facebook, Instagram, WhatsApp, Viber, Messenger, Gmail, Skype, and YouTube. This data set will be shared openly with the research community to foster new studies and allow reproduction of the results presented. (https://www.dropbox.com/s/9tihcj9wx2sia1t/Dataset.7z?dl=0 (accessed on 4 October 2021))

The rest of this paper is organized as follows: In Section 2, the related work is reviewed. Section 3 describes the methodology of the proposed classification system. Section 4 presents the experimental results and discussions. Section 5 provides concluding remarks.

2. Related Work

In the literature, two types of research methods have dominated traffic classification: statistical methods and neural networks.

Statistical classification is a technique that exploits the statistical characteristics of network traffic flow to perform traffic classification. Features, such as packet length, packet duration, packet inter-arrival time, and traffic flow idle time, are used in this method. To perform an actual classification based on statistical features, classifiers, specifically ML algorithms, are used. Conti et al. [9] performed mobile user actions classification based on packet sizes and their order. Saltaformaggio et al. [5] developed NetScope to detect user activities based on inspecting statistical features obtained from internet protocol (IP) headers. Taylor et al. [2,10] used features, such as packet lengths and statistical properties of flows, to train support vector classifier (SVCs) and random forest classifiers to perform mobile app classification. Pathmaperuma et al. [4] identified user activities performed on mobile apps using statistical features generated from frame length, inter-arrival time, and direction leaked from encrypted traffic. Wang et al. [3] computed 20 statistical features from frame size and inter-arrival time to train an RF classifier to perform app identification. Zhang et al. [11] proposed a scheme by combining supervised and unsupervised machine learning techniques to classify apps. Twenty unidirectional flow statistical features related to packet size, packets, inter-packet time, and bytes were extracted and used to train the proposed classifier. A classification method was proposed by Draper-Gil et al. [12] with only time-related flow features on both regular encrypted traffic and protocol encapsulated traffic.

Many works have applied neural networks in network traffic analysis such as malware classification [13], anomaly detection [14], DDoS attacks detection [15], and intrusion detection [16,17]. In [18], Wang et al. proposed an end-to-end encrypted traffic classification approach with 1D CNN to detect traffic types such as streaming, VoIP, and file transfer. Lopez-Martin et al. [19] proposed using a recurrent neural network (RNN) combined with a CNN for IoT traffic classification. In [20], Aceto et al. performed traffic classification in encrypted network flow using DL techniques. Wang [21] proposed a stacked auto encoder (SAE)-based method to detect network protocols. The results showed that Wang's approach worked well on the applications of feature learning, protocol classification, anomalous protocol detection, and unknown protocol identification. In [22], Lotfollahi et al. proposed a framework to perform traffic characterization and application identification using DL, embedding an SAE and CNN to classify network traffic.

Recent studies have explored the use of CNNs to perform classification by converting network traffic flows into images. Wang et al. [13] converted traffic to 2D images and then applied 2D CNN to classify the traffic images achieving the goal of malware classification. Ma et al. [8] proposed a CNN model that predicts large-scale, network-wide traffic speed. In this work, spatiotemporal traffic dynamics were converted to images describing the time and space relations of traffic flow via a 2D time–space matrix. Zhou et al. [6] proposed a classification algorithm Min–Max Normalization (MMN) CNN that processes traffic data and maps them into gray images that are input into a CNN to detect 12 types of traffic categories such as mail, game, and multimedia. Tavakoli [23] presented the Seq2Image method to perform human genomic sequence classification converting genome sequences to images and then using a 2D CNN to classify the created images of sequences. Shapira et al. [7] presented the FlowPic approach to transform flow data into an image and then used image classification DL techniques, a CNN, to identify the flow category and application in use. Kim et al. [24–26] presented the NetViewer approach that detects and visualizes attacks and anomalous traffic by passively monitoring packet headers. In these works, multiple pieces of traffic data are represented as different colors of an image. Image processing techniques are applied to generated images to analyze the network traffic. He et al. [27] proposed an image-based method that converts the first few non-zero payload sizes of session to gray images and uses a 1D CNN to perform the classification.

The works in the literature primarily focus on classifying previously trained traffic, while none has considered performing network traffic analysis accurately in the presence of noise generated by unknown traffic, even though this would be a typical situation in a real-world scenario. Thus, this work aimed to advance the state-of-the-art by identifying previously trained fine-grained in-app activities accurately as well as by detecting and classifying previously untrained in-app activities as unknown data.

To detect unknown in-app activities, the function needs to reject the classification label for those inputs belonging to classes never exposed during training. An output-based rejection technique is proposed in this work that leverages additional information from the deep learning model output such as the SoftMax probabilities of each class. Usually in classification tasks, the neuron with the highest probability will be chosen and the corresponding class label is assigned. In this work, a two-stage approach was used to check if the neuron with the highest probability satisfied a pre-set threshold value. Based on this, the known and unknown instances were classified. However, setting this threshold is challenging, as setting it too high increases false negatives, whereas setting it too low increases false positives. To test the impact of the threshold value on the model's performance, a range of threshold values were selected, and tests were performed.

3. Proposed Methodology

This paper proposes a CNN-based method that transfers network traffic flows into images to identify in-app activities while detecting unknown data. The method contains two main procedures. The first is to convert network traffic into images that represent side channel data of a network flow as a 2D image. The second is to apply image classification DL techniques to the generated images and perform traffic classification of previously trained and untrained apps' traffic flows.

3.1. Data Collection

For this research, a data set was created that consisted of network traffic from in-app activities of eight popular mobile apps, namely, Facebook, Instagram, YouTube, Viber, WhatsApp, Gmail, Skype, and Messenger. To obtain the ground truth, network traffic generated after executing each activity was collected separately, and the network trace was labeled with the name of the activity performed. Each app was run separately, thus limiting the presence of background traffic. To generate a sufficient number of traffic flows for inference, each activity was repeated four times and captured traffic was saved as .pcap files. The number of in-app activities considered in each app and number of samples obtained after segmenting the traffic into 1, 0.5, 0.2, and 0.1 s time intervals are presented in Table 1. The 92 in-app activities considered in this research are given in Appendix A.

We monitored the network passively and traffic was captured without connecting to the wireless network to which the target user's mobile device was connected. The traffic transmitting within the wireless network was sniffed using Airmon-ng and Airodump-ng sniffing tools from the Aircrack-ng suite [28]. Default setting of network adapters allow only to capture packets that are sent to them. To capture all traffic, we set the network adapter to its monitor mode. The experimental testbed used to sniff traffic is shown in Figure 2. The internet access to the smartphone was provided over a wireless connection via a router. The smartphone was connected to the wireless network exclusively to avoid interference from other sources. The network adapter (Alfa AWUS036NHA) was plugged to the laptop (Toshiba PORTEGE Z30-C), which was used to capture the network traffic.

Table 1. Features of the collected data set.

App	Category	No. of in-App Activities	Number of Samples			
			1 s	0.5 s	0.2 s	0.1 s
Facebook	Social networking	22	9477	19,944	55,283	110,588
Instagram	Photo and video	20	2907	6818	25,529	81,505
Gmail	Productivity	5	561	1036	2361	5514
Messenger	Social networking	10	2199	6061	18,514	43,113
Skype	Social networking	8	5436	15,703	69,884	211,981
Viber	Social networking	9	434	690	1207	1875
WhatsApp	Social networking	9	273	436	873	1458
YouTube	Photo and video	9	5161	14,501	58,173	207,013
Total 8 apps		92	26,448	65,189	231,824	663,047

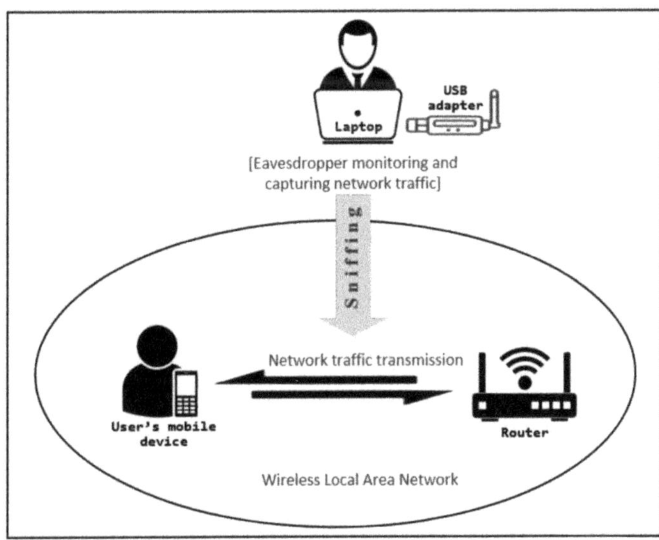

Figure 2. Experimental testbed for data collection.

3.2. Network Traffic to Image

Pre-processing needed to be applied on the captured traffic flows prior to mapping traffic data into images. The following pre-processing steps were performed on the obtained data set in this work.

(a)   Sanitization

The three main frame types used in WLAN (IEEE 802.11) are data, control, and management frames. Only the data frames are used for data transmission. The presence of the other two types may hinder the process of analysis; therefore, the control and management frames were eliminated, and only data frames were processed further. There were data frames that did not carry data. Keeping these frames would cause bias in training the CNN [29]. Therefore, null data frames were also eliminated at this stage.

(b)   Normalization

Normalization is an important step in data pre-processing to avoid having different scales of feature vectors and, thus, improves integrity of data. The data set contained feature values in different scales that may lead to obtaining biased results. To normalize the feature values, the standard scaler [30] was used, which normalizes each feature by removing its mean and scaling to unit variance. This equalizes the importance of all features and allows DL methods to converge faster. For each original value X with a mean μ and standard deviation σ, its normalized value X' can be determined from (1):

$$X' = (X - \mu)/\sigma \tag{1}$$

(c)   Segmentation

During in-app activity detection, it is not possible to guarantee that the entire transaction of an activity can be observed. There can be situations where the eavesdropper starts to capture the traffic while the user is already performing an activity. In these instances, the eavesdropper can capture only a part of the traffic flow instead of the entire flow transaction. To perform in-app activity detection even by observing only a part of an activity related traffic, time windows are used to divide the traffic flows into segments. A thorough analysis of the sensitivity of this approach was performed by conducting experiments with different time window sizes: 1, 0.5, 0.2, and 0.1 s as discussed in Section 4.1.

Following network traffic pre-processing, each segmented traffic sample was converted into an image, where features and corresponding feature values were represented by pixels and pixel intensities, respectively. In this work two sets of data images were generated that used different color schemes. These were grayscale and red, green, and blue (RGB). If the segmented traffic sample, S, has n number of frames and m number of data values, then the value of mth data at nth frame can be represented as A_{nm}. The segment S is represented as matrix (2):

$$S = \begin{bmatrix} A_{11} & \cdots & A_{1m} \\ \vdots & \ddots & \vdots \\ A_{n1} & \cdots & A_{nm} \end{bmatrix} \tag{2}$$

The three main side channel data considered in this work were frame length, data length, and inter arrival time. Segment S comprised values of these side channel data. Based on these data, individual vectors were created. The division can be expressed as matrix (3):

$$X_i = [A_{1i}, A_{2i}, \ldots A_{ni}] \tag{3}$$

where n is the number of frames and X_i represents all data values of the ith side channel data. The number of vectors that are created depends on the number of side channel data. In this research, three side channel data were considered; thus, three vectors were created. When employing a grayscale model, these three vectors combined to form one vector. Whereas in the RGB model, these were passed together as three separate vectors.

If the image size is very small, then it cannot be sent through the required number of convolutional layers, because after each layer, the size is reduced. Therefore, resizing was applied to match with the respective pre-defined input image size.

Each element in the matrix is treated as a pixel with the grayscale value of the pixel, where the color intensity is proportional to the matrix value. Figure 1 shows three in-app activities' traffic flows in grayscale format. The input image dimensions were h × w × c, where h × w were the dimensions of the image, and c was the number of channels. The image dimension wass constructed according to the number of features in the data set. The dimensions of the images were set to 28 × 28 following empirical tests. Grayscale images had one channel. Hence, the input image size of grayscale images was resized to 28 × 28 × 1 (784). Data fed into CNN must be uniform. The data that were less than these pre-defined image sizes were zero-padded, and data that were more than the pre-defined image sizes were truncated to match the respective pre-defined size.

3.3. Image Classification Using CNN

The CNN architecture proposed to identify in-app activities was composed of four main parts: model input, traffic feature extraction, prediction, and model output as shown in Figure 3.

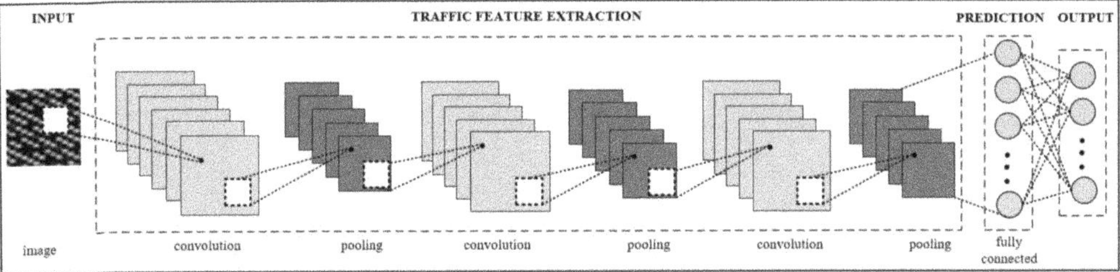

Figure 3. Proposed CNN architecture to identify in-app activities. The traffic flow data transferred into pixelized images are used as the input to the CNN model with input dimension of 28 × 28.

The input layer was then connected to a combination of convolution and pooling layers to learn image features. In this proposed architecture, three convolutional layers were used, where each convolution layer was followed by a pooling layer that selects the most important features from its receptive region and reduces the number of parameters required to train the model. Pooling layers reduce the size of the output through the convolutional process and cancel noise. In our model, the max-pooling function was used, which outputs the maximum value in a rectangular neighborhood of the previous layer.

The fully connected layers at the end of CNN concatenated the output of the convolution layer into a dense vector. This was passed to the model prediction phase.

The dense vector was then transformed into model outputs through a fully connected layer, which had 92 outputs indicating the number of classes (in-app activities) used to classify the input images. This output layer was a SoftMax layer, which output a K-dimensional probability distribution vector of values in the range of 0–1, where K is the number of classes (K = 92). Each node value represents a class score. The class with the highest score was selected and the corresponding class label was assigned.

The hyperparameters, such as the number of convolutional layers, number of fully connected layers, number of filters for each hidden layer, filter and stride size, and activation functions, were selected through comprehensive tests involving numerous parameter combinations. The depth of the CNN should be neither too large nor too small [31], and thus it is able to learn complex relations while maintaining model's convergence. To determine the suitable value for the model's depth, different values from small to large were assigned to test the CNN model until the best model was found. The rationale behind using this architecture was based on the experiments we conducted in which we discovered that this architecture was the best fit model for our problem.

3.4. Unknown In-App Data Detection

In a real-world setting, data traffic captured contains both previously known and unknown traffic flows. Previously known traffic is related to the in-app activities considered during model training whereas unknown traffic relates to the in-app activities not considered during model training. A major challenge to the robustness of the classifier's performance comes from previously unknown/unseen traffic. Identifying previously untrained in-app activities using the proposed method is one of the key contributions of this work.

When the traffic flows are converted into images and input to the CNN model, they pass through hidden layers and reach the output layer. A SoftMax layer is added to the end of the CNN, which converts the output values into a probability distribution. Model's output layer has nodes that is equal to the number of classes. The SoftMax layer provides probabilities for each class label in the interval (0, 1). Usually for a given input sample, one of the classes will have a higher probability value than the rest of the classes. In normal traffic classification tasks, a class is assigned to a data point based on the highest probability. However, in this work instead of making the class with the highest probability be the final classification, a threshold approach is used to determine if the test sample is a known or unknown instance. Figure 4 shows the technique used to detect noise (unknown in-app data) generated by previously unknown traffic. *Pmax* denotes the node with the highest predicted class probability. The decision for converting the predicted probability into a class label is dominated by a parameter known as the *threshold*. If *Pmax* < *threshold*, then the test sample is labeled as an unknown instance. Threshold set on a positive class determines whether the test sample belongs to one of the trained classes, which translates into a pre-trained in-app activity or not.

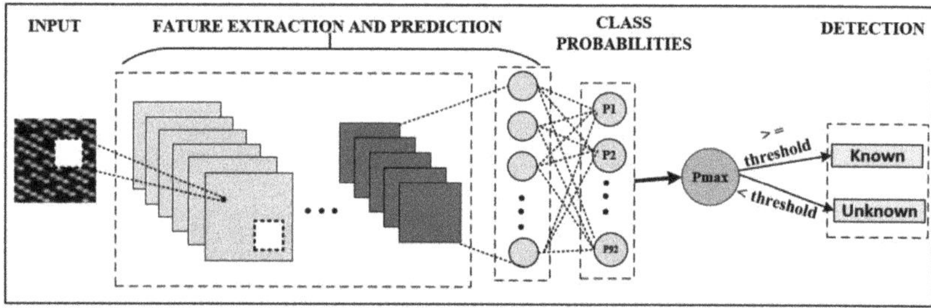

Figure 4. Framework to detect unknown in-app activity data.

To examine the impact of the threshold on model's performance, a range of threshold values were selected and tested. Setting a threshold too high results in an increase in false negatives, while setting it too low leads to an increase in false positive. Therefore, setting a balanced threshold value is challenging. In this work, a threshold value that contributes to achieving the highest classification accuracy was utilized, which was obtained empirically as 0.97.

4. Experimental Results and Discussion

The performance evaluation of the proposed model is presented in this section. When evaluating the proposed model, the following two factors are considered to measure the performance.

- Ability to detect previously trained in-app activities correctly;
- Ability to detect previously untrained/unknown in-app activities as noise data.

The CNN models were constructed, trained, and tested by Keras 2.5.0 with TensorFlow 2.5.0 running at the back end. Experiments were conducted using 92 in-app activities from the Facebook, Instagram, YouTube, Viber, WhatsApp, Gmail, Skype and Messenger apps (see Table 1). Eighty percent of the total samples were dedicated for training, and remaining 20% for validation.

Accuracy was used to evaluate the model performance, which is computed as follows:

$$\text{Accuracy of known data} = \frac{\text{True Positives} + \text{True Negatives}}{\text{Total no. of known instances}} \tag{4}$$

$$\text{Accuracy of unknown data} = \frac{\text{True Negatives}}{\text{Total no. of unknown instances}} \tag{5}$$

4.1. Model Analysis

The proposed CNN architecture comprises of seven layers excluding the input and output layers (as depicted in Figure 3). The three convolutional layers had 256, 128, and 64 numbers of filters in each layer, respectively. The fully connected layer had 128 nodes. Tanh activation function is applied to the output of every convolutional and fully connected layer. To reduce overfitting, the dropout technique is used to prevent complex co-adaptations on the training data [7]. In this architecture, the Adam optimizer and categorical cross entropy loss function were used.

The model performance varied with different image sizes as inputs according to our tests. When the image size was too small, there was a sign of learning degradation. When the image size was too large, then the extraction phase took much longer time. Thus, we selected 28×28 pixel image size, which helped to reduce the run-time and memory consumption while improving detection rate in all our experiments.

In addition to grayscale images, RGB images were also created from the collected network traffic data set. This was done to observe the performance of in-app activity classification when in-app data is converted to 3D images. Instead of considering the entire flow of an activity, we used the segment-based approach to perform the classification. Four different time windows were used to divide the traffic flows into segments: 1 s, 0.5 s, 0.2 s, and 0.1 s. Table 2 presents the training and validation accuracies, and time needed to train and test the models.

Table 2. Classification performance of grayscale and RGB models.

	Window Sizes							
	1 s		0.5 s		0.2 s		0.1 s	
	Gray	RGB	Gray	RGB	Gray	RGB	Gray	RGB
Training accuracy (%)	97	84	98	84	96	83	95	81
Validation accuracy (%)	83	71	88	76	92	80	86	74
Training and testing time	8 min	15 min	19 min	27 min	1 h 9 min	1 h 20 min	2 h 30 min	2 h 48 min

From the experimental results, it can be observed that the accuracy values obtained for the grayscale models are higher than those recorded for the RGB models for all time windows tested. This is because RGB models suffer from overfitting due to the presence of large number of features resulting from the three color channels. Therefore, grayscale images are selected as inputs to avoid false classification and complexities for the rest of the experiments.

Validation accuracy was highest when the time window size was 0.2 s. The reason is when traffic traces are segmented into a smaller window size, we were able to obtain plenty of samples. Training the model with large of samples contributed positively towards the classification accuracy. However, this was not true when the window size was further reduced. Even though the sample size increased when traffic traces were segmented into a smaller window size such as 0.1 s, the classification accuracy decreased. This was because the number of frames contained in such smaller window segments is less. When frames were considered individually or when the number of frames was insignificant, they contained very little information to perform the classification.

With the decrease in window size, leading to an increase in the number of samples, the results present that long times are needed to pre-process, train, and test the models. From the Table 2 data, it can be observed that all grayscale models achieved an average accuracy of 87%. This shows that the proposed model can identify fine-grained in-app activities even by observing only a small subset of an activity's traffic.

4.2. Unknown In-App Data (Noise) Detection

In noise detection tests we use leave-one-out approach where two data sets were created, namely training and noise data sets. From the eight apps considered in the experiments, each time an app was singled out and used to create the noise data set. The remaining seven apps were used to create the train data set. While the training data set was used to train the model, the noise data set (with in-app activities unknown to the trained model) was input to the trained model to determine its ability to detect unknown data. The result of this experiment is shown in Figure 5.

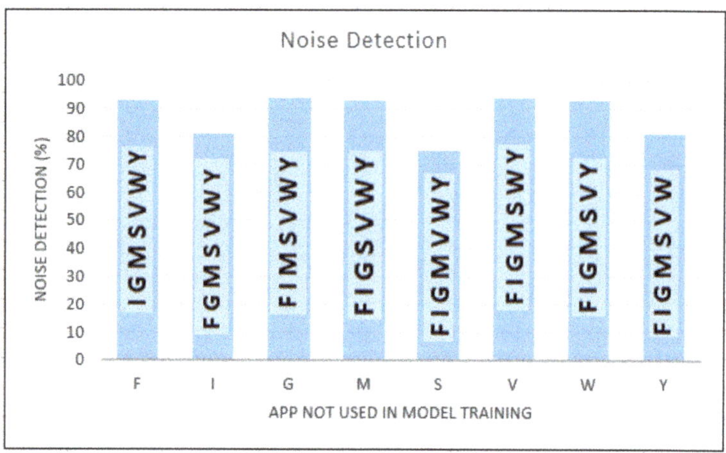

Figure 5. Noise detection rates of unknown apps.

In Figure 5, F, I, G, M, S, V, W, and Y denote Facebook, Instagram, Gmail, Messenger, Skype, Viber, WhatsApp, and YouTube, respectively. On each vertical bar in Figure 5, the apps used to train the model at each instance are denoted. The app labelled on the X axis is the app in the noise data set.

Let's us explain how to interpret Figure 5. The first vertical bar in Figure 5 shows that a CNN model is trained with the in-app activity data from the following 7 apps: I, G, M, S, V, W and Y. The in-app activity data from F is kept away from the training process. During the testing, the in-app activity data from F is used to determine the robustness of trained CNN model. As shown in the first bar in Figure 5, the trained CNN model successfully identified more than 90% of the in-app data from F as unknown data.

As shown in Figure 5, the proposed method achieved 75% or more in detecting noise with average accuracy of 88%. Data from Gmail and Viber applications were detected with 94% accuracy. However, the proposed model couldn't distinguish data from Skype with the trained data only 75% of the Skype traffic got correctly detected as unknown traffic. The remining 25% was misclassified as in-app activities that belong to the training data set.

To understand the nature of the misclassified traffic, further analysis was performed on the apps to which the unknown traffic got classified. The results are presented in Figure 6 in percentage values (%). For example, only 8% of the data from F is classified as known data (see first bar in Figure 5). The distribution of this 8% of the misclassified data from F is shown in the first row in Figure 6. Majority of this data (35%) are assigned to S (Skype).

Test app		F	I	G	M	S	V	W	Y
	F		15	2	13	35	2	1	32
	I	33		2	10	28	2	1	24
	G	33	8		9	24	1	1	24
	M	36	10	2		26	1	1	24
	S	42	12	2	11		2	1	30
	V	32	10	2	9	24		1	22
	W	31	10	2	9	24	1		23
	Y	40	12	2	12	31	2	1	

Figure 6. Confusion matrix (% values).

Looking at Figure 6, the majority of the misclassified data was assigned to Facebook. We observed that there was a high correlation among all the considered apps with Facebook, which contributed to the misclassification. For Instagram, Gmail, Messenger, Viber, and WhatsApp, the second and third highest misclassifications came from Skype and YouTube, respectively. This can also be seen by plotting the correlation between these apps (see Figure 7).

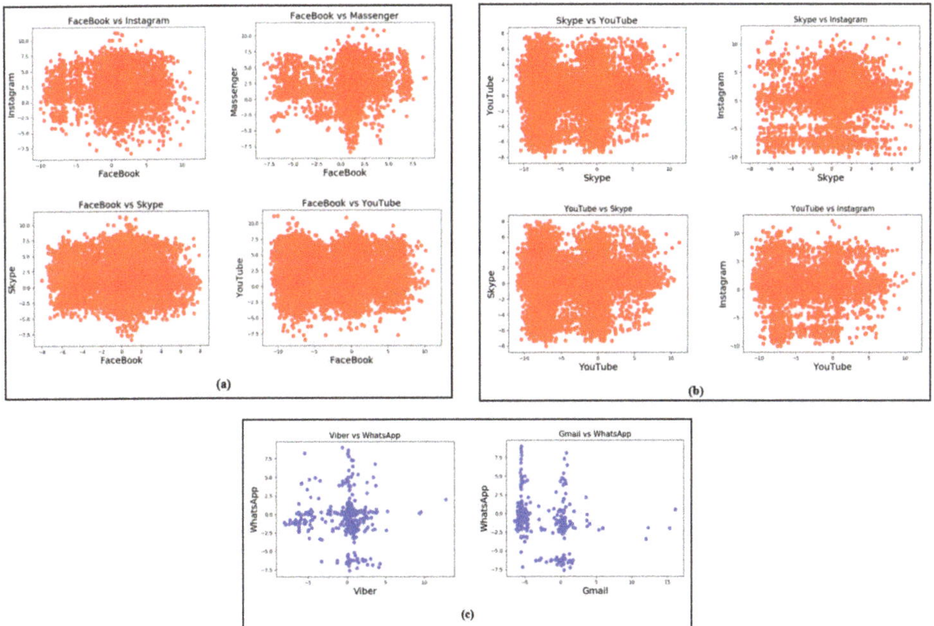

Figure 7. Correlations among different apps. (**a**) correlation of Facebook with Instagram, Messenger, Skype and YouTube. (**b**) correlation of Skype vs YouTube and Instagram with YouTube and Skype. (**c**) correlation of WhatsApp with Viber and Gmail.

Figure 7a shows the correlation plots of Facebook with four different apps. Figure 7b shows the correlation plots of Skype and YouTube with two different apps. In Figure 7a,b the points are closely packed, which means the strength of the correlation was high. Apps that are highly correlated with each other have similar in-app activities with similar behavior.

Looking at the W column in Figure 5, the least amount of misclassification resulted from WhatsApp, which was 1% for all the test apps. This was due to the least correlation between W and other apps as shown in Figure 7c. It can be observed that the points are distributed loosely which means the considered apps are only slightly correlated to each other. The behavior of Facebook, Instagram, Gmail, Messenger, Skype, Viber, and YouTube in-app activities was not similar to WhatsApp's in-app activities, which makes them less correlated to each other.

To obtain a better insight into the misclassifications occurred, further analysis of the in-app activities to which the test apps got misclassified was conducted. Table 3 lists the six in-app activities with the highest misclassification percentage against each of the test apps, where the in-app activities were coded as follows:

Utuv—uploading a video on YouTube;
Skvc—having a video call on Skype;
Msgac—having an audio call on Messenger;
Skv—sending a video on Skype;
Indvc—having a video chat on Instagram Direct;
Fblive—uploading a live video on Facebook;
Skac—having an audio call on Skype;
Fbpv—posting a video on Facebook wall
Utwv—watching a video on YouTube.

Table 3. Misclassified in-app activities with misclassification percentage (%) values.

F	I	G	M	S	V	W	Y
Utuv—27	Utuv—21	Utuv—21	Utuv—21	Utuv—26	Utuv—19	Utuv—19	Fblive—20
Skvc—13	Fblive—15	Fblive—16	Fblive—17	Fblive—22	Fblive—15	Fblive—14	Skvc—11
Msgac—9	Skvc—9	Skvc—8	Skac—9	Msgac—8	Skvc—7	Skvc—8	Msgac—8
Skv—9	Skac—8	Skac—7	Skvc—7	Fbpv—6	Skac—7	Skac—7	Skac—8
Skac—8	Msgac—7	Msgac—6	Fbpv—6	Indvc—6	Msgac—6	Msgac—6	Skv—8
Indvc—6	Skv—7	Fbpv—6	Skv—5	Utwv—4	Skv—6	Skv—5	Fbpv—7

The highest percentage of misclassification was recorded from Utuv on Facebook and Skype. When Facebook was input to the model as the unknown/noise app, 27% of its traffic was misclassified as Utuv. When creating the Facebook data set, activities such as posting a video on wall, uploading a live video were considered, which are very similar to Utuv on YouTube. Therefore, having such similarity in the in-app activities caused the misclassification. When Skype was considered as the unknown app, 26% of its traffic was also misclassified as Utuv. Similarly, when creating the Skype data set activities such as sending a video message, engaging in a video call were considered. Again, having such similar in-app activities to Utuv has caused the reported misclassification.

While most of the Facebook traffic was misclassified as Utuv, which is in fact an activity from YouTube; most of the YouTube traffic got misclassified as Fblive, an activity from Facebook. Both Utuv and Fblive activities are related to video uploading and thus have a high correlation between each other due to their similarity in behaviour, resulting in the misclassification.

4.3. Performance Comparison

Even though in this work a CNN model was used to perform the traffic classifications, other types of neural networks could also be employed for this purpose, such as a Deep Neural Network (DNN). This section reports on the performance comparison between the DNN and CNN models when they are used in our tests.

In both models, the output types were in-app activities. But the format of the input was different: in CNN, images were provided as the input whereas in DNN, statistical features related to the traffic flows constituted the input. Although the same data set was used for both DNN and CNN models to perform tests, we employed the network architecture that resulted in the highest accuracy for each model instead of using the same network architecture across the board for a fair comparison. This was needed as the input format was different in both cases, which had a direct influence on accuracy performance. For CNN, the network architecture proposed in Section 4.1 was used. For the DNN model, four hidden layers were used with 1024, 512, 256 and 128 nodes in each layer. Tanh activation function was used for all layers expect for the output layer which used the SoftMax layer. The 48 statistical features used in [4] were utilized as the input of DNN. Figure 8 shows the training and validation accuracies obtained at 0.5 s and 0.2 s time window sizes when both models were used to perform in-app activity classification.

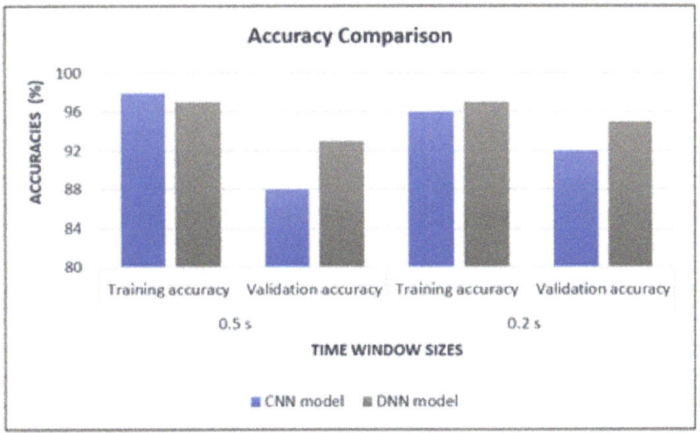

Figure 8. Accuracy comparison of the CNN and DNN models.

From Figure 8, it is noted that the DNN model has recorded the highest accuracy values compared to the CNN model in all categories. But when looked closely at the comparison at each time window size, the difference in accuracy values is maximum 5%. Significantly, all the accuracies of the CNN model are at 88% or above. Therefore, it can be concluded that both models can accurately detect previously trained in-app activities.

To compare the detection of unknown in-app data, it can be observed from Figure 9 that the CNN model has outperformed the DNN model at all the instances when different noise test traffic data sets (Test app) were applied. In both models, Gmail and Viber reported the highest noise detection rates. Compared to the DNN model, when the CNN model is employed, there is an increment of 19%, 18%, and 12% noise detection rates when YouTube, Messenger, and Skype are input as test app, respectively.

Even though the ability to detect previously trained in-app activities correctly by the CNN model is slightly weaker than that of DNN, its ability to detect previously un-trained/unknown in-app activities as noise data is much stronger than that of DNN. Therefore, when the overall performance is considered, the proposed CNN model out-performs DNN. This is because when traffic flows are converted to images, the model can apply image processing techniques to reveal interesting properties of the traffic. As such, applying the proposed CNN model on the input traffic images allows for extracting and selecting salient features that enable the model to learn to differentiate trained and unknown traffic.

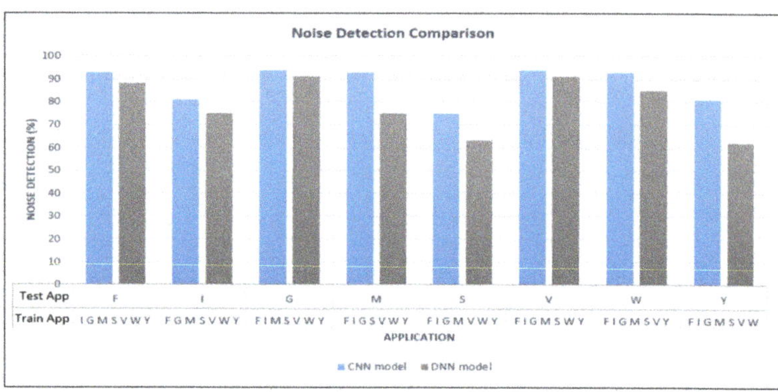

Figure 9. Noise detection comparison of the CNN and DNN models.

5. Conclusions

In this paper, a novel approach was introduced for encrypted Internet traffic classification, both for identifying known and unknown traffic and categorizing in-app activity type, based only on frame's size and time related information. User actions identified through analysing network traffic can be used in forensic investigations and security incident analysis, to improve correlation of events. Profiling users based on their in-app activities is also useful for marketing or intelligence purposes. Deep Learning obviates the need to select features by a domain expert as it automatically selects features through training, making it a desirable approach when new classes constantly emerge, and patterns of old classes evolve. Performance of the proposed CNN based method that learns traffic as images was compared with DNN that uses statistical features. The results demonstrate that the proposed CNN model has outperformed DNN when overall performance is considered. Moreover, a windowing approach was used to perform classification by observing a short time window of a flow instead of the entire session. Even though this is significantly a harder task as there is less information in partial encrypted traffic flow compared to the entire flow, our model was able to identify in-app activities with an accuracy of 92% even by observing the traffic only for 0.2 s. The novel approach of using a threshold on the confidence values exploits the model's output layer to identify in-app activities while removing noise traffic generated by untrained in-app activities with an average accuracy of 88%.

Author Contributions: Conceptualization, M.H.P., Y.R., S.D. and A.K.; methodology, M.H.P., Y.R. and S.D.; software, M.H.P.; validation, M.H.P.; formal analysis, M.H.P.; investigation, M.H.P., Y.R. and S.D.; data curation M.H.P.; writing—original draft preparation M.H.P.; writing—review and editing, Y.R. and S.D.; visualization, M.H.P.; supervision, Y.R., S.D. and A.K. All authors have read and agreed to the published version of the manuscript.

Funding: This research received no external funding.

Data Availability Statement: This data set is shared openly with the research community to foster new studies and allow reproduction of the results presented. (https://www.dropbox.com/s/9tihcj9wx2sia1t/Dataset.7z?dl=0 (accessed on 10 December 2021)).

Acknowledgments: Madushi H. Pathmaperuma would like to express her thanks to the Loughborough University, UK, for supporting her research. Safak Dogan would like to acknowledge the Engineering and Physical Sciences Research Council (EPSRC), UK, for their support of his work (EP/W00366X/1).

Conflicts of Interest: The authors declare no conflict of interest.

Appendix A

Table A1. The 92 in-app activities considered in this research.

Application	Category	Fine Grained Activity
Facebook	Social networking	Post an image on wall
		Post a video on wall
		Post a long text on wall
		Post a short text on wall
		Post a feeling on wall
		Post a check in on wall
		Post a live video on wall
		Comment a short text on a post
		Comment a long text on a post
		Comment a post with a sticker
		Comment a post with an image
		Add an image to a story
		Add a video to a story
		Add a text to a story
		Share an image to the wall
		Share a video to the wall
		Share a text to the wall
		Like an image
		Like a video
		Like a comment
		Send a friend request
		Watch Facebook video
Instagram Instagram- Direct	Photo and Video	Add an image to a story
		Add a video to a story
		Add a text to a story
		Like an image
		Like a video
		Like a comment
		Comment a short text on a post
		Comment a long text on a post
		Post an image on feed
		Post a video on feed
		Follow a friend
		Follow back a friend
		Send a message to a story
		Watch Instagram video
		Send a long text message
		Send a short text message
		Send a voice recording message
		Send an image
		Like a message
		Video chat
YouTube	Photo and Video	Watch a video
		Like a video
		Dislike a video

Table A1. *Cont.*

Application	Category	Fine Grained Activity
YouTube	Photo and Video	Comment a long text on a video
		Comment a short text on a video
		Like a comment
		Dislike a comment
		Upload a video
		Subscribe a channel
Skype	Social networking	Video call
		Audio call
		Send a long text message
		Send a short text message
		Send voice recording
		Send an image
		Send a video
		Send a file
Gmail	Productivity	Send short text email
		Send long text email
		Send an image
		Send a video
		Send a file attachment
Messenger	Social networking	Video call
		Audio call
		Send a ling text message
		Send a short text message
		Send a voice recording
		Send an image
		Send a video
		Add an image to a story
		Add a video to a story
		Add a text to a story
WhatsApp	Social networking	Send long text message
		Send short text message
		Video call
		Audio call
		Send voice recording
		Send an image
		Send a video
		Send location
		Send contact
Viber	Social networking	Send long text message
		Send short text message
		Video call
		Audio call
		Send voice recording
		Send an image
		Send a video
		Send location
		Send contact

References

1. Rezaei, S.; Liu, X. Deep learning for encrypted traffic classification: An overview. *IEEE Commun. Mag.* **2019**, *57*, 76–81. [CrossRef]
2. Taylor, V.F.; Spolaor, R.; Conti, M.; Martinovic, I. Appscanner: Automatic fingerprinting of smartphone apps from encrypted network traffic. In Proceedings of the 2016 IEEE European Symposium on Security and Privacy (EuroS&P), Saarbruecken, Germany, 21–24 March 2016; IEEE: Manhattan, NY, USA, 2016; pp. 439–454.
3. Wang, Q.; Yahyavi, A.; Kemme, B.; He, W. I know what you did on your smartphone: Inferring app usage over encrypted data traffic. In Proceedings of the 2015 IEEE Conference on Communications and Network Security (CNS), Florence, Italy, 28–30 September 2015; IEEE: Manhattan, NY, USA, 2015; pp. 433–441.
4. Pathmaperuma, M.H.; Rahulamathavan, Y.; Dogan, S.; Kondoz, A.M. In-app activity recognition from Wi-Fi encrypted traffic. In *Science and Information Conference*; Springer: Cham, Swizerland, 2020; pp. 685–697.
5. Saltaformaggio, B.; Choi, H.; Johnson, K.; Kwon, Y.; Zhang, Q.; Zhang, X.; Xu, D.; Qian, J. Eavesdropping on fine-grained user activities within smartphone apps over encrypted network traffic. In Proceedings of the 10th USENIX Workshop on Offensive Technologies (WOOT 16), Austin, TX, USA, 8–9 August 2016.
6. Zhou, H.; Wang, Y.; Lei, X.; Liu, Y. A method of improved CNN traffic classification. In Proceedings of the 2017 13th International Conference on Computational Intelligence and Security (CIS), Hong Kong, China, 15–18 December 2017; IEEE: Manhattan, NY, USA, 2017; pp. 177–181.
7. Shapira, T.; Shavitt, Y. Flowpic: Encrypted internet traffic classification is as easy as image recognition. In Proceedings of the IEEE INFOCOM 2019-IEEE Conference on Computer Communications Workshops (INFOCOM WKSHPS), Paris, France, 29 April–2 May 2019; IEEE: Manhattan, NY, USA, 2019; pp. 680–687.
8. Ma, X.; Dai, Z.; He, Z.; Ma, J.; Wang, Y.; Wang, Y. Learning traffic as images: A deep convolutional neural network for large-scale transportation network speed prediction. *Sensors* **2017**, *17*, 818. [CrossRef]
9. Conti, M.; Mancini, L.V.; Spolaor, R.; Verde, N.V. Analyzing android encrypted network traffic to identify user actions. *IEEE Trans. Inf. Forensics Secur.* **2015**, *11*, 114–125. [CrossRef]
10. Taylor, V.F.; Spolaor, R.; Conti, M.; Martinovic, I. Robust smartphone app identification via encrypted network traffic analysis. *IEEE Trans. Inf. Forensics Secur.* **2017**, *13*, 63–78. [CrossRef]
11. Zhang, J.; Chen, X.; Xiang, Y.; Zhou, W.; Wu, J. Robust network traffic classification. *IEEE/ACM Trans. Netw.* **2014**, *23*, 1257–1270. [CrossRef]
12. Draper-Gil, G.; Lashkari, A.H.; Mamun, M.S.I.; Ghorbani, A.A. Characterization of encrypted and vpn traffic using time-related. In Proceedings of the 2nd International Conference on Information Systems Security and Privacy (ICISSP), Fredericton, NB, Canada, 19 February 2016; pp. 407–414.
13. Wang, W.; Zhu, M.; Zeng, X.; Ye, X.; Sheng, Y. Malware traffic classification using convolutional neural network for representation learning. In Proceedings of the 2017 International Conference on Information Networking (ICOIN), Da Nang, Vietnam, 11–13 January 2017; IEEE: Manhattan, NY, USA, 2017; pp. 712–717.
14. Tang, T.A.; Mhamdi, L.; McLernon, D.; Zaidi, S.A.R.; Ghogho, M. Deep learning approach for network intrusion detection in software defined networking. In Proceedings of the 2016 International Conference on Wireless Networks and Mobile Communications (WINCOM), Fez, Morocco, 26–29 October 2016; IEEE: Manhattan, NY, USA, 2016; pp. 258–263.
15. Niyaz, Q.; Sun, W.; Javaid, A.Y. A deep learning based DDoS detection system in software-defined networking (SDN). *arXiv* **2016**, arXiv:1611.07400. [CrossRef]
16. Mirsky, Y.; Doitshman, T.; Elovici, Y.; Shabtai, A. Kitsune: An ensemble of autoencoders for online network intrusion detection. *arXiv* **2018**, arXiv:1802.09089.
17. Shone, N.; Ngoc, T.N.; Phai, V.D.; Shi, Q. A deep learning approach to network intrusion detection. *IEEE Trans. Emerg. Top. Comput. Intell.* **2018**, *2*, 41–50. [CrossRef]
18. Wang, W.; Zhu, M.; Wang, J.; Zeng, X.; Yang, Z. End-to-end encrypted traffic classification with one-dimensional convolution neural networks. In Proceedings of the 2017 IEEE International Conference on Intelligence and Security Informatics (ISI), Beijing, China, 22–24 July 2017; IEEE: Manhattan, NY, USA, 2017; pp. 43–48.
19. Lopez-Martin, M.; Carro, B.; Sanchez-Esguevillas, A.; Lloret, J. Network traffic classifier with convolutional and recurrent neural networks for Internet of Things. *IEEE Access* **2017**, *5*, 18042–18050. [CrossRef]
20. Aceto, G.; Ciuonzo, D.; Montieri, A.; Pescapè, A. MIMETIC: Mobile encrypted traffic classification using multimodal deep learning. *Comput. Netw.* **2019**, *165*, 106944. [CrossRef]
21. Wang, Z. The applications of deep learning on traffic identification. *BlackHat USA* **2015**, *24*, 1–10.
22. Lotfollahi, M.; Siavoshani, M.J.; Zade, R.S.H.; Saberian, M. Deep packet: A novel approach for encrypted traffic classification using deep learning. *Soft Comput.* **2020**, *24*, 1999–2012. [CrossRef]
23. Tavakoli, N. Seq2image: Sequence analysis using visualization and deep convolutional neural network. In Proceedings of the 2020 IEEE 44th Annual Computers, Software, and Applications Conference (COMPSAC), Madrid, Spain, 13–17 July 2020; IEEE: Manhattan, NY, USA, 2020; pp. 1332–1337.
24. Kim, S.S.; Reddy, A.N. A study of analyzing network traffic as images in real-time. In Proceedings of the IEEE 24th Annual Joint Conference of the IEEE Computer and Communications Societies, Miami, FL, USA, 13–17 March 2005; IEEE: Manhattan, NY, USA, 2005; Volume 3, pp. 2056–2067.

25. Kim, S.S.; Reddy, A.N. Image-based anomaly detection technique: Algorithm, implementation and effectiveness. *IEEE J. Sel. Areas Commun.* **2006**, *24*, 1942–1954. [CrossRef]
26. Kim, S.S.; Reddy, A.N. Modeling network traffic as images. In Proceedings of the IEEE International Conference on Communications, 2005, ICC 2005, Seoul, Korea, 16–20 May 2005; IEEE: Manhattan, NY, USA, 2005; Volume 1, pp. 168–172.
27. He, Y.; Li, W. Image-based encrypted traffic classification with convolution neural networks. In Proceedings of the 2020 IEEE Fifth International Conference on Data Science in Cyberspace (DSC), Hong Kong, China, 27–30 July 2020; IEEE: Manhattan, NY, USA, 2020; pp. 271–278.
28. Aircrack-ng. Available online: https://www.aircrack-ng.org/ (accessed on 11 July 2021).
29. Taheri, S.; Salem, M.; Yuan, J.S. Leveraging image representation of network traffic data and transfer learning in botnet detection. *Big Data Cogn. Comput.* **2018**, *2*, 37. [CrossRef]
30. Sklearn.preprocessing. StandardScaler. 2021. Available online: https://scikit-learn.org/stable/modules/generated/sklearn.preprocessing.StandardScaler.html?highlight=standardscaler#sklearn.preprocessing.StandardScaler (accessed on 15 July 2021).
31. Lv, Y.; Duan, Y.; Kang, W.; Li, Z.; Wang, F.Y. Traffic flow prediction with big data: A deep learning approach. *IEEE Trans. Intell. Transp. Syst.* **2014**, *16*, 865–873. [CrossRef]

future internet

MDPI

Review

A Review of Blockchain Technology Applications in Ambient Assisted Living

Alexandru-Ioan Florea, Ionut Anghel * and Tudor Cioara

Computer Science Department, Technical University of Cluj-Napoca, Memorandumului 28, 400114 Cluj-Napoca, Romania; alexandru.florea@staff.utcluj.ro (A.-I.F.); tudor.cioara@cs.utcluj.ro (T.C.)
* Correspondence: ionut.anghel@cs.utcluj.ro

Abstract: The adoption of remote assisted care was accelerated by the COVID-19 pandemic. This type of system acquires data from various sensors, runs analytics to understand people's activities, behavior, and living problems, and disseminates information with healthcare stakeholders to support timely follow-up and intervention. Blockchain technology may offer good technical solutions for tackling Internet of Things monitoring, data management, interventions, and privacy concerns in ambient assisted living applications. Even though the integration of blockchain technology with assisted care is still at the beginning, it has the potential to change the health and care processes through a secure transfer of patient data, better integration of care services, or by increasing coordination and awareness across the continuum of care. The motivation of this paper is to systematically review and organize these elements according to the main problems addressed. To the best of our knowledge, there are no studies conducted that address the solutions for integrating blockchain technology with ambient assisted living systems. To conduct the review, we have followed the Preferred Reporting Items for Systematic Reviews and Meta-Analyses (PRISMA) methodology with clear criteria for including and excluding papers, allowing the reader to effortlessly gain insights into the current state-of-the-art research in the field. The results highlight the advantages and open issues that would require increased attention from the research community in the coming years. As for directions for further research, we have identified data sharing and integration of care paths with blockchain, storage, and transactional costs, personalization of data disclosure paths, interoperability with legacy care systems, legal issues, and digital rights management.

Keywords: ambient assisted living; blockchain; security and privacy; IoT blockchain integration; decentralization

Citation: Florea, A.-I.; Anghel, I.; Cioara, T. A Review of Blockchain Technology Applications in Ambient Assisted Living. *Future Internet* **2022**, *14*, 150. https://doi.org/10.3390/fi14050150

Academic Editor: Christoph Stach

Received: 19 April 2022
Accepted: 11 May 2022
Published: 12 May 2022

Publisher's Note: MDPI stays neutral with regard to jurisdictional claims in published maps and institutional affiliations.

1. Introduction

Ambient assisted living (AAL) is a research field that aims to bring smartness to our everyday environments by acquiring data from various sensors, understanding people's activities, behavior, and living problems, and deciding on proactive interventions to support the management of identified issues (see Figure 1) [1]. With the advance in sensing technologies and the prevalence of miniaturized, affordable Internet of Things (IoT) sensor applications have been developed to improve the beneficiary's quality of life and support personalized care [2]. A wider range of proof of concept applications for various use scenarios along with associated technologies can be found in the literature, such as fall detection systems [3], cognitive decline management [4], personalized care [5], remote follow-up [6], nutrition management [7], medication review [8] and well-being management [9].

Having more IoT devices and sensors associated with living environments leads to collecting patient data that must be shared among multiple parties on different sides [10]: validators, processors, healthcare stakeholders, etc. Nowadays, the ambient assisted living systems move the data collected in cloud systems (see Figure 1), where the potentially unlimited computation resources help in dealing with analytics and decision making [11].

Using decentralized distributed ledger solutions will allow multiple nodes to host the same set of encrypted data in multiple care systems that are hosted in different places and kept up to date with the actual state and data of the system [12]. Additionally, most of the collected data are rather sensitive and personal data of vulnerable people; thus, security and privacy have always been issues to deal with [13]. They constitute a barrier between vulnerable people and assistive technologies and prevent the adoption and good use of existing software solutions in the field [14]. People may become digitally vulnerable as data theft, fraud, and the unauthorized use of personal, medical, and financial information are often not even known by the victims [15].

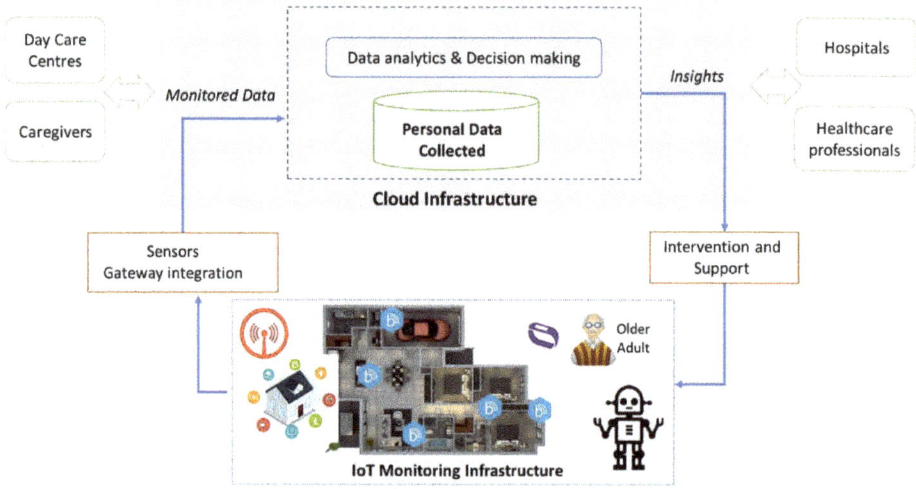

Figure 1. Conceptual architecture of a cloud-based AAL system.

The privacy and security problems are critical for data-driven assisted living applications and IoT networks such as the Internet of Medical Things [14]. Data ownership and elimination of potential breaches are objectives for keeping the data, and the system secured [16]. However, because of the lack of precise specifications, even ordinary procedures might result in security breaches [17]. There is a strong need to make such applications transparent, immutable, and distributed [18]. In general, in the discussions concerning privacy and security, how consumers understand privacy is key [19]. People are more inclined to value decentralized solutions for their capacity to safeguard their privacy goals [20]. On top of technical privacy issues, lately, personal details (i.e., used to identify a person) have become one of the most valuable commodities [21]. This information might be as basic as a name or identification number, or it can be more sensitive, such as medical or behavioral data [22]. As the world becomes more digitized, internet activity is increasingly recorded, often without the user's knowledge or agreement, constituting a barrier to ambient technologies adoption.

Blockchain technology is seen as a good solution for tackling IoT monitoring, data management, interventions, and privacy concerns [23] in ambient assisted living applications [24]. Stakeholders from the ambient assisted and care fields are interested in integrating blockchain technologies into their systems to benefit from improved security, privacy, and data ownership (see Figure 2) [24]. Conventional ambient assisted living solutions use centralized cloud-based models focused on structuring data rather than privacy, ownership, and decentralization. The adoption of blockchain technology can change this landscape [25]. In a blockchain-driven assistive living application, the users will join a blockchain network, and asymmetric encryption solutions will enforce the security of data sharing [26]. The IoT devices deployed in the user environment can be joined with smart contracts to automatically generate and sign transactions and forward them to the blockchain to be stored immutably [27]. The generated transactions are aggregated in blocks disseminated in the network and will be mined in the future blocks. To change a value in a block, the entire history of previously linked blocks needs to be rehashed, requiring a lot of computational power not being feasible these days [28]. To prove the ownership of the data, the IoT device provides a signature of the transaction, which is also useful for authentication and validation [29]. The updates are stored in chained blocks, taking advantage of the technology's properties such as reliability, availability, immutability, and consensus [30]. The blockchain enforces the provenance of data by a linked list of nodes; thus, data can be traced back by iteration of the chain [31]. In addition, securely storing the sensor's data and respecting the personal data regulations is difficult considering the perspective of the domain [32]. So, using a decentralized, user-centric approach regarding data privacy can address security and data ownership problems in developing ambient assisted living applications.

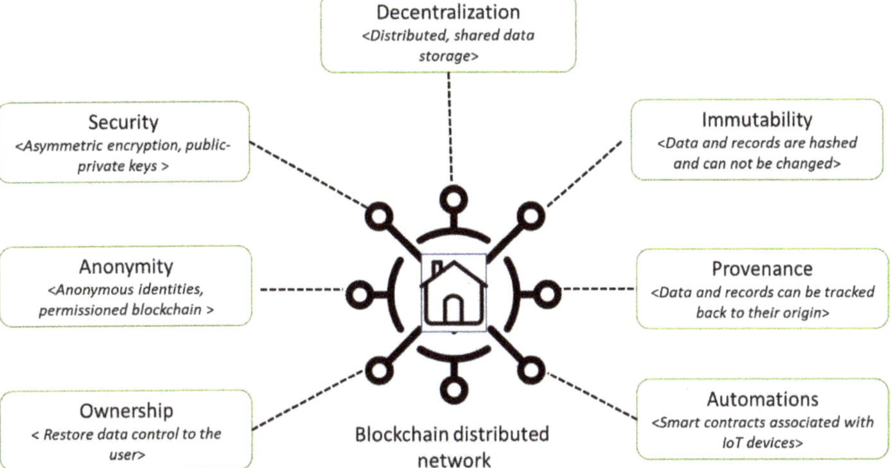

Figure 2. Features of blockchain technology desired for an ambient assisted living system.

Even though the research field is still at the beginning, relevant studies in the literature can be found. The motivation of this paper is to systematically review and organize them according to the research problems they address. To the best of our knowledge, there are no reviews conducted on solutions for integrating blockchain technology with ambient assisted living systems. To conduct this review, we have defined a search methodology with clear criteria for including or excluding papers, set up a reference interval, and focused on current important databases. We have included in the survey 87 papers on blockchain and ambient assisted living systems that were reviewed and organized. Thus, a reader would effortlessly gain insights into the current state-of-the-art research in the field. Nevertheless, there are still many gaps and open issues that would require increased attention from the research community in the coming years, such as data sharing and integration of care paths with blockchain, storage, and transactional costs, personalization of data disclosure paths, interoperability with legacy systems, legal issues, digital rights management, etc.

The remainder of the article is organized following the Introduction, Methods, Results, and Discussion (IMRAD) structure: Section 2 presents the methodology and methods used in conducting the literature review; Section 3 illustrates the results by describing and organizing the most relevant research works; Section 4 presents a discussion on the survey findings, and Section 5 draws the conclusions.

2. Materials and Methods

In carrying out our study, we have used the PRISMA methodology that defines the guidelines for conducting systematic reviews, which is widely accepted by most Web of Science (WoS) journals for organizing review-type articles [33]. More specifically, we have selected the "PRISMA 2020 flow diagram for new systematic reviews which included searches of databases and registers only" variant that features four main phases: articles identification, screening, eligibility, and inclusion. The goal of our systematic study is to create an overview of the domain of blockchain and IoT applications for ambient assisted living and to construct a snapshot of the state-of-the-art works for general or specific topics in this domain. This approach will also allow identifying current hot trends, future research directions, and research gaps.

The first step in our research study was to clearly define search strategy in terms of research questions, keywords, or key phrases to cover the study targeted topic of blockchain, IoT, and AAL applications. The following research questions have been selected for our study:

- Identify use-cases for blockchain and IoT applications in AAL;
- Identify applications, techniques, and tools developed for this domain;
- Highlight the challenges and limitations of blockchain and IoT in AAL;
- Find what are the main open research directions to be tackled

This led to the definition of the following main search keywords to be used in the next stage of the study:

- Blockchain IoT healthcare;
- Blockchain and Ambient Assisted Living;
- Blockchain and Active Assisted Living;
- Blockchain and Ambient Intelligence
- Blockchain and remote care;
- Blockchain data ownership in health care;
- Blockchain data sharing and analytics in health care;
- Blockchain and IoT security and privacy in health care

Using the above, the second step of applying the PRISMA methodology was to select the scientific databases for the search process. In this context, we have selected Web of Science as the main database for our study since it is the most comprehensive scientific database widely recognized for including high-quality conference and journal articles from the most important publishers (MDPI, IEEE, Elsevier, ACM, Springer, Wiley, etc.). Using the WoS database allowed us to focus our search on a single platform while receiving results from articles from multiple publishers. To conduct the search, we have used the Clarivate WoS web platform [34]. As a search method in this platform, we have selected the Topic type because it covers the key information from the WoS indexed research articles: title, abstract, author, keywords, and Keywords Plus. The search keywords have been transformed into search strings in the platform, e.g., "blockchain" AND "ambient intelligence".

Figure 3 illustrates the PRISMA 2020 flow diagram used to identify the articles that were included in the review. In the PRISMA identification phase, after aggregating the search results, we have identified 491 articles matching our search criteria. We have refined this set of articles and removed duplicate records (19 items), resulting in 472 records to be included in the Screening phase. In this phase, we have defined specific inclusion criteria for our study to further filter the results, thus, removing 312 records. Similarly, to further narrow the set of records in the Eligibility phase, we defined several exclusion criteria that helped us drop another 73 records. Both criteria are presented in Table 1.

Finally, in the inclusion phase, we obtained 48 records to be considered in the study for an in-depth analysis of the presented work, concepts, approaches, and solutions for blockchain in AAL.

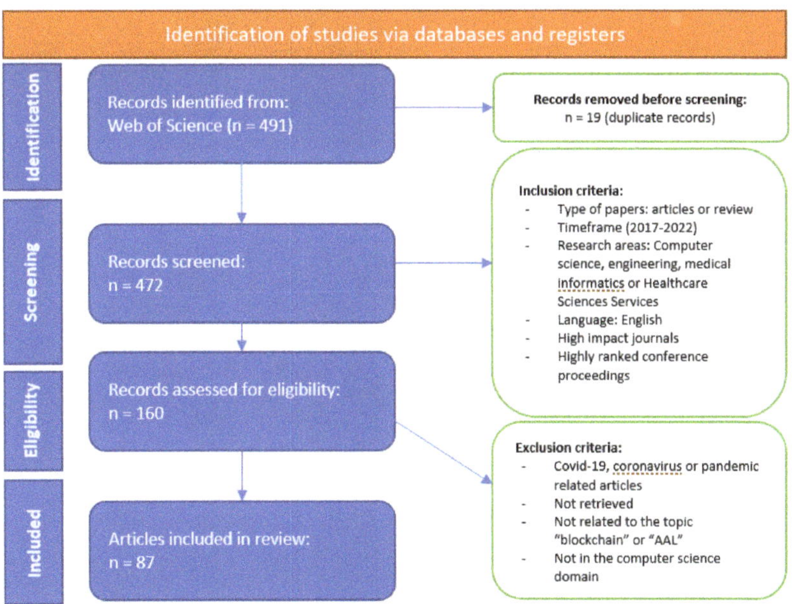

Figure 3. PRISMA 2020 flow diagram for the current study.

Table 1. Criteria for including and excluding articles in the study.

Screening Phase Inclusion Criteria	Eligibility Phase Exclusion Criteria
Type of papers: articles or review	Not retrieved
Timeframe (2017–2022)	Not related to the topics: blockchain, IoT, and AAL
Research areas: Computer science, engineering, medical informatics, or Healthcare Sciences Services	COVID-19, coronavirus, or pandemic-related articles
Language: English	Not in the computer science domain
High impact journals	
Highly ranked conference proceedings	

Figure 4 presents the included papers distribution per publishing year. It can be noticed that most of the research around the studied topics has been accelerated from 2020 onwards.

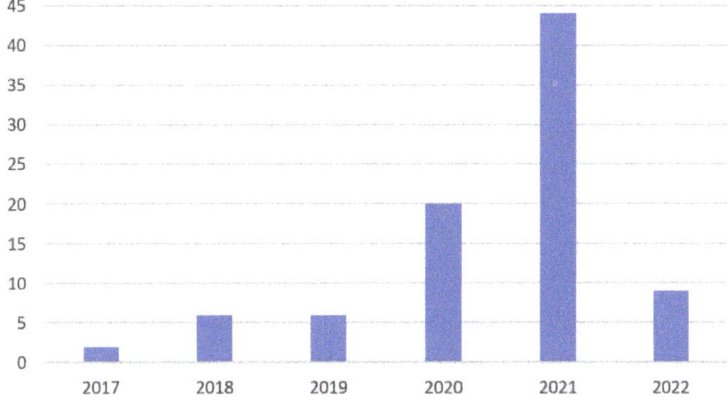

Figure 4. Results distribution by year.

Figure 5 shows the distribution of the selected articles using the journal/conference publisher as criteria. As it can be seen in the figure, all major highly rated publishers (Elsevier, IEEE, MDPI, and Springer) have shown interest in the blockchain and AAL research direction, 80% of the selected papers being published under one of the four.

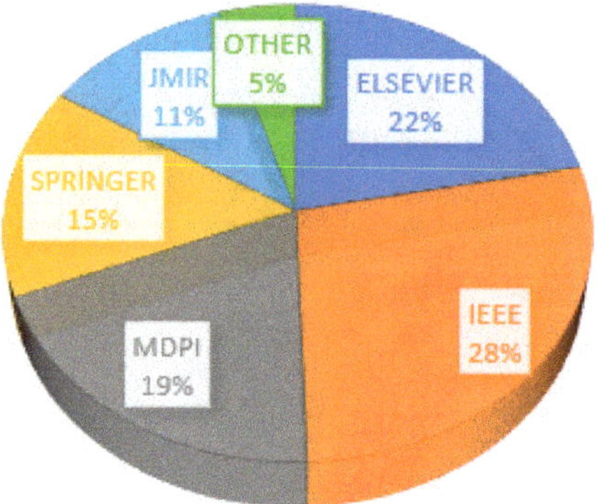

Figure 5. Results distribution by publisher.

As per the types of papers included in the study, in Figure 6, we illustrate the main categories of the analyzed papers, with an emphasis on article types.

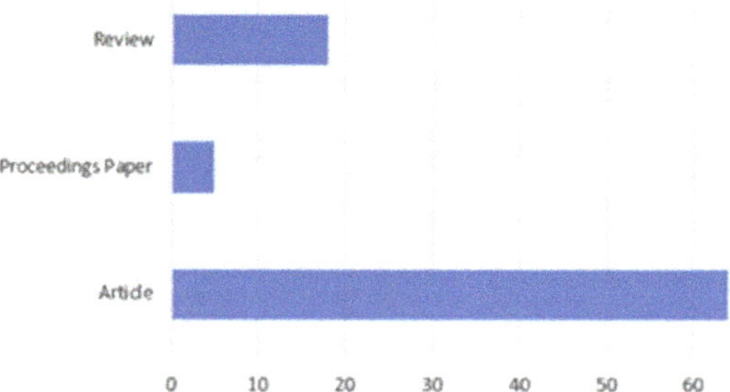

Figure 6. Selected paper types.

Figure 7 shows the distribution of the included papers per journal and conference proceeding highlighting that more research related to the study domain has been published in IEEE Access and Sensors MDPI journals.

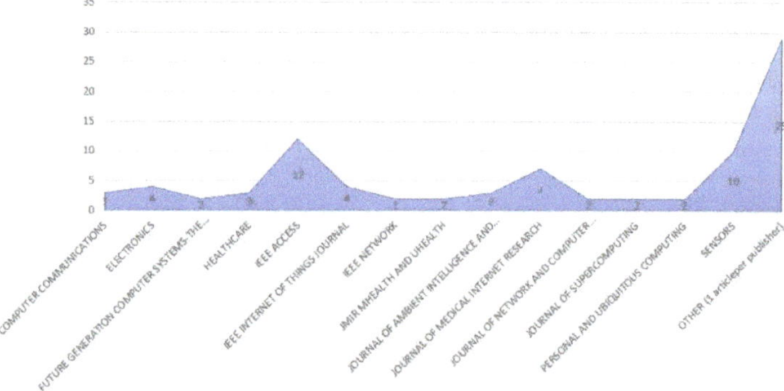

Figure 7. Journal and conference proceedings comparison.

Table 2 summarizes the query results as the number of records together with the number of items included in the study per each category.

Table 2. Results overview.

Search Keywords/Query Phrase	Identified Records	Selected Items (after Removing Duplicates)	Articles References
Blockchain and IoT	342	38	[13–15,17–22,24–32,35–54]
Blockchain and Ambient Assisted Living	15	5	[55–59]
Blockchain and Active Assisted Living	2	1	[60]
Blockchain and Ambient Intelligence	7	1	[61]
Blockchain and remote care	39	22	[62–83]
Blockchain data ownership in healthcare	15	6	[84–89]
Blockchain data sharing and analytics in healthcare	16	7	[10,16,90–94]
Blockchain and IoT security and privacy in healthcare	55	7	[95–101]

3. Results

After identifying and selecting the relevant papers using the defined criteria, we have conducted a qualitative analysis to identify blockchain applications and use cases in ambient assisted living systems and the associated challenges and limitations. Most of the literature on blockchain application in healthcare focuses on the health aspects, such as the management of electronic health records, and only a few relevant papers were found on addressing aspects of patient care at home using ambient assisted living systems. Nevertheless, most of the identified papers are very recent, mostly beyond 2020, showing that blockchain technology usage in ambient assisted living is a fast-emerging field of research that will gain a lot of attention in the near future.

We have organized the reviewed papers on the basis of the most important aspects of ambient assisted living reported in the literature to which blockchain may bring significant improvements: (1) monitoring, timely follow-up, and intervention of patients or older adults living at home using IoT devices; (2) decentralized data storage to avoid single point of failure, data manipulation issues, and mistrust; and (3) privacy and security aspects of cross-continuum of care.

3.1. Patient Monitoring and Intervention

Integrating IoT with blockchain technology is used to develop decentralized ambient monitoring and intervention infrastructures using IoT devices (see Table 3). The data provided by the IoT devices can be stored on the blockchain as transactions and replicated in all the nodes of the network. The blockchain can offer an efficient environment for disseminating IoT-acquired patient data in a secure way to all relevant healthcare stakeholders [63]. Blockchain can reinforce trust and address problems related to limited access to healthcare [67].

Table 3. Blockchain usage benefits of IoT monitoring in AAL systems.

Ambient Assisted Living Use Case	Blockchain Usage	References
IoT-based monitoring and intervention	Reinforce trust, provenance tracking	[63,67,72]
	Remote monitoring and telemedicine	[63,65,67–69,94]
	Patient data-efficient dissemination and interoperability	[17,18,57,64,71,84]
	Personalized care services	[17,65,70,94]
	Automation using smart contracts	[17,28,66,75]

The integration of IoT with blockchain may contribute to the relief of pressure on sanitary systems while simultaneously providing tailored care services to enhance people's quality of life [65,94]. The growing geriatric population with chronic medical conditions increased the adoption of IoT devices for remote at-home monitoring [68] and telemedicine [69]. These have become even more evident during the COVID-19 pandemic [63,67]. An IoT taxonomy relevant for ambient assisted living systems is provided in [16,47]. Five categories have been identified: sensor-based, resource-based, communication-based, software-based, and security-based methods. Blockchain has promising features for developing data-flow architectures that integrate the monitoring devices and assure patient data-efficient dissemination to relevant healthcare stakeholders [57,64]. The marriage of blockchain with Internet of Things technology supports the paradigm shift towards preventive and personalized care systems [17,70]. Although blockchain and IoT adoption in ambient assisted living systems is still in its early stages, it can address flaws in the care processes [18] and close some communication gaps, assuring better interoperability [71,84]. An investigation into the development of smart ambulances is presented in [19]. The blockchain can be used to increase interoperability and efficiency of information exchanges with the hospital for timely intervention in an emergency department [72].

Blockchain is a good choice for establishing a decentralized, self-contained IoT system deployed in older adults' homes [75]. In [28], the authors propose an IoT based on blockchain integration architecture, a rich–thin client IoT technique for addressing the challenges associated with the restricted IoT capacities when adopting blockchain in remote monitoring and healthcare processes. In this context, the smart contracts can assure a seamless and automatic solution platform connecting a range of IoT devices relevant for remote follow-up [66]. Smart contracts are a crucial feature of blockchain technology that enables it to be used in a variety of ambient assisted living systems [14,85]. However, the smart contract concept, its operation, and how it can be used in ambient assisted living are still poorly understood [48].

Nevertheless, the main barrier to integrating IoT monitoring devices with blockchain in the context of ambient assisted living systems is scalability [62]. The researchers of [24] provide an overview of blockchain technology and explore prominent consensus methods utilized in the healthcare processes. However, as the authors pointed out in [36], there are issues to be solved, including scalability and standardization. Research has been conducted to improve the scalability of IoT to blockchain integration [73]. This is a relevant aspect of ambient monitoring and assistive services [37]. In [44], the authors propose a blockchain framework that is described as more accurate, precise, and efficient than other popular methods of storing and accessing patient records among personnel, medical stakeholders, and facilities. In [18], the authors discuss the evolution of healthcare, identifying the research gaps such as the relocation of care from hospital to home and ambient assisted care [5] that we consider to be a relevant use case for joining Wireless Body Area Networks with blockchain [74]. Improved scalability of a permissioned blockchain framework has been described in [51] using Hyperledger Fabric as an infrastructure for the blockchain network. The authors of [31] investigated a composite scalability concept which can be seen differently depending on the grade of innovation we want to achieve. The notion of blockchain scalability is discussed, including techniques and ideas for increasing core blockchain functionality and blockchain-based applications in domains such as remote care [76,81]. Blockchain and fog computing is being used in care IoT to provide safe and trustworthy transactions [95]. An Extended Signature-Based Encryption technique is proposed for healthcare IoT device authentication, as well as authorization [52]. The authors claim the suggested architecture and algorithm may offer safe transaction and transmission. Finally, [54] looks at blockchain to IoT systems and how to make them more scalable. On-chain and off-chain methodologies are contrasted, and suggestions are made to help designers create scalable blockchain-based IoT medical systems.

3.2. Decentralized Data Management

As healthcare processes become more digitalized, issues concerning safe storage, ownership, and sharing of patient personal health records and related medical data have arisen [10], and they can be addressed by using blockchain technology (Table 4). To address some of the issues mentioned above, patient data can be stored in the cloud [59], and security policies can be applied via smart contracts [86]. Increasing the amount of time that vulnerable persons can spend at home alone and how data acquisition systems can efficiently share data using blockchain are presented in [87]. Removing some of the barriers to adopting electronic patient records in management platforms through a blockchain is presented in [88]. The data collected from IoT devices can be stored in an Interplanetary File System (IPFS) storage while data access and interactions are managed through smart contracts executed on a blockchain [96]. In [12], the authors identify four possible research directions for blockchain technology in the healthcare domain: scalability, privacy and security, digital currency management, and cross-chain technology.

Table 4. Blockchain usage benefits data management in AAL systems.

Ambient Assisted Living Use Case	Blockchain Usage	References
Decentralized patient data management	Safe, decentralized storage of data	[42,59,77,82,91]
	Data sharing and smart contracts	[25,38,39,87,88]
	Data ownership	[12,14,25,26,49,83]
	Health and care processes integration	[89,90]
	Data analytics	[40,42,43,46]

A blockchain-based architecture is presented in [38] to enable a distributed patient data sharing and smart-contract-based web service automation while not compromising the security and privacy of the system. To tackle the limitations of cloud-based systems, such as single point of failure, the use of decentralized data storage systems is proposed [42]. An essential insight into the possibilities of blockchain technology, particularly in the health sector, is provided in [25]. The authors discuss the reasons for using smart contract technology in healthcare and the prospects of using smart contracts in health records and sharing processes, medical testing, pharmaceutical manufacturers, big data, machine learning, security, and privacy, among other areas. The significant barriers to the use of blockchain technologies, along with scalability and storage conditions, are interoperability with legacy systems [49].

In [39], a data-sharing strategy based on the IPFS was developed, which not only increases the availability of data but also decreases data redundancy among the many stakeholders of the care ecosystem. In [14], the use of blockchain technology is discussed regarding the use of IoT in the remote monitoring of patients. Blockchain is seen as a good technology for facilitating the implementation of the internet of medical things, which refers to the interconnectedness of devices and sensors in the healthcare domain that collects real-time data [89]. The challenge with this data is that it is typically stored in a centralized location, which creates a single point of failure and raises privacy and security concerns [90], and a consortium blockchain network with smart contracts and interplanetary file systems can provide secure storage and transmission of data [77]. Important research directions in joining blockchain and ambient assisted living services are scalability, response time, blockchains interoperability, privacy, and ownership of data [83].

Big data, artificial intelligence, and distributed ledger technology, among other technologies, are blurring the barriers between the physical and digital worlds. Blockchain is a new technology and needs the development of more efficient and scalable strategies for incorporating it into the existing healthcare processes [43]. An innovative blockchain-based business model for health and care systems is described in [40]. It places the patient at the heart of the paradigm and may be used in any business situation with a set of user incentive criteria. As indicated in [42], blockchain technology may be utilized to enhance

IoT-driven care systems and tackle different difficulties. It offers a two-stage architectural solution for integrating IoT with blockchain using dew and cloudlet computing. In [46], a blockchain framework is proposed that allows data owners to create preferred access controls for electronic patient records. A two-chain architecture is used to store access controls as well as data transactions and employs a clustering strategy to handle the real growth difficulties associated with distributed ledgers.

Even though, as listed above, blockchain technology brings potential benefits for data management applications to ambient assisted living systems, there are challenges that limit its adoption [30,37]. Services built on fuzzy systems and blockchain technology are proposed in [49] to provide a behavior-driven intuitive security measure for healthcare IoT environments and networks based on blockchain. In [85], the authors explore different methods for assisting in medication usage. Blockchain can be used to store and disseminate information concerning adverse responses to prescription pharmaceuticals [97]. In [26], various scenarios are presented in the form of an analysis that verifies the key aspects of establishing, verifying, and changing people's identities. It presents various blockchain identity verification solutions available on the market built on top of public or private blockchains. Finally, a significant amount of time might be saved if patient characteristics are disseminated among all relevant stakeholders across the care continuum [91], illustrating how distributed ledgers and blockchain technology might be used for AAL systems to support decentralized data management [82].

3.3. Security and Privacy

Blockchain technology usage in ambient assisted living systems brings benefits (see Table 5) for addressing flaws and vulnerabilities, such as security flaws in smart IoT devices [41], trust and security, as well as the interoperability of such systems with legacy applications [20]. It may also circumvent the restrictions of client/server architectures in cloud-based ambient assisted living applications because of its scattered peer-to-peer nature [13].

Table 5. Blockchain usage benefits for the security and privacy of AAL systems.

Ambient Assisted Living Use Case	Blockchain Usage	References
AAL system security and privacy	GDPR compliant applications	[32,35,78,79]
	Informed consent management	[14,21,27,35,50,56,58,60,99–101]
	Data privacy and identity management	[13,15,29,45,50]
	Security and confidentiality	[20,22,27,30,50,80,93]

A thorough literature review of GDPR-compliant blockchains was conducted in [32]. The essential GDPR for blockchains can be broken down into six categories that include data removal and modification, security by design, data controller and data processor obligations, consent management [35], data processing norms and lawfulness, and geographical reach. In [79], new research paths are proposed, such as the adoption of private blockchains to support the implementation of ambient assisted living systems. The authors of [41] examined recent breakthroughs in IoT-based healthcare procedures identifying critical challenges for systems development such as security, privacy, and authentication. In [43], blockchain technology was utilized to address such challenges to build a more efficient and dependable care system. Additionally, blockchain provides relevant features for ambient assisted living systems, such as data tampering and service failures [78]. Utilizing a distributed ledger, data might be visible to all users and, therefore, would allow for data integrity and provenance tracking verification [92].

In [35], a thorough assessment of the implementation of blockchain technology in the sphere of consent is presented, as well as privacy and data management. The consent of the patient is an important topic in the field of ambient assisted living systems [50]. In [60], a study was conducted on better techniques to manage informed consent so that data access is not abused and personal data protection regulations are respected. A good platform for obtaining informed consent from both a patient and a proxy (in the case of patients who have no discernment or cannot make decisions for themselves) is the Hyperledger fabric network blockchain [27] with smart contracts. Blockchain platforms may combine several roles and stakeholders in the care system, such as institutions that regulate access to personal data, data consumers, research institutions such as universities, and devices and technologies that acquire data [98]. A consent management framework that incorporates Distributed Ledger Technologies is presented in [21]. The platform offers an onion-layered secure way of transmitting sensitive data and a better way of accessing management methods. IPFS can be used for sharing files in a safe, transparent, and decentralized way in ambient assisted living systems [14]. Similar solutions based on another type of blockchains can be found in the literature [99], but they are not focused directly on consent, even if they can be interpreted as access to data itself [100].

Lately, differential privacy has emerged as perhaps the most successful privacy guarding solution for IoT medical and care infrastructure [13]. Some experimental findings and confidentiality proofs that demonstrate a particular suggested protocol that has a reasonable computational cost, as well as security safeguards for digital healthcare transactions, are presented in [45]. For decentralized and trustworthy healthcare data interactions, smart contracts and Elliptic Curve Encryption can be employed. One goal of ambient assisted living data protection protocol is to be resistant to a variety of threats and to have reasonable operating and computing capabilities [15]. A blockchain solution using a zero-knowledge-based authentication architecture to tackle privacy issues is presented in [29]. The architecture authenticates devices without revealing any information about the identity of the user. The paper also introduces the ZKNimble cipher, which is suitable to be used by devices that do not benefit from a good processing power.

According to [27], the ambient assisted living and care systems should deliver and share patient data through a secure transfer to ensure the confidentiality of data [93]. This was made feasible using a blockchain-based approach to the system's architectural design [55], while work still needs to be carried out on interoperability. It is explained in [20] how many IoT applications in healthcare are no different from those in any other area, and the research should concentrate on the industry's unique requirements, such as good levels of privacy and security. In this context, the number of blockchain-based applications in healthcare has increased lately, but the domain highly demands interdisciplinary studies [30]. A blockchain classification for IoT applications is provided in [47]. The authors explore the most prevalent blockchain systems for healthcare.

In [50], a blockchain-based solution is proposed for managing private data using Hyperledger Fabric and Caliper and can be used in various domains such as healthcare. The core benefits of blockchain technology for such systems are immutability, traceability, and transparency [80]. In [22], a decentralized and scalable architecture is presented supporting device access, authentication, as well as data security. A novel authentication protocol has been devised and constructed on Physical Unclonable Functions cryptographic primitives. This makes it practically difficult to predict the key values of the protocol because of the randomness provided by the physical architecture of the protocol. The system suggested in [53] enables medical officials to authenticate data received by a common wearable device with a verification error of less than 1% and a price compared with fewer as being much cheaper for one hour of observing the activity. A decentralized specific ring-based authorization method, as well as an authentication scheme and patient's records anonymity algorithms, are provided to increase the proposed system's security [58]. It allows for decentralized automated identity management, privacy, and security [61]. Finally, in [15], the authors use blockchain to protect patients' anonymity and privacy from several potential threats while enabling important institutions to interact with one another. Only authorized users have access to the genuine identities, addresses, as well as medical data of patients in the proposed system [101]. Authorities should use blockchain to tackle the issue of cyber-attacks tampering with sensor data [56].

4. Discussion

As the population of the world is aging, societal challenges will need to be faced, especially about the delivery of care, which needs to be improved, and new care system paths need to be designed. At the same time, the development of IoT sensors and technology such as blockchain can ease this process by the implementation of ambient assisted living services which aim at moving the care from hospitals and care centers to home. The integration of sensors in older adults or patients' homes to enable remote follow-up and care is seen as key in delaying.

The ambient assisted living systems address many of the concerns of patients in this transition towards remote care and personalized interventions, such as (1) time-consuming process for healthcare professionals caused by the lack of accurate monitoring and follow-up support, (2) patient data-sharing gaps across the care continuum, (3) not having a proper care support network in place to reduce patient anxiety or worries; (4) patients and family caregivers lacking sufficient knowledge and skills to optimize self-care; (5) patient difficulties in adherence to postdischarge instructions, e.g., medication usage or behavioral changes.

Despite advantages brought to the care process, the ambient assisted living solutions have a rather limited adoption mostly because of the problems related to IoT sensors integration, data sharing, trust, ethical considerations, data confidentiality, privacy, etc. (see Table 6). As shown by the qualitative review conducted, blockchain technology can play a significant role in addressing some of the concerns related to the ambient assisted living services adoption, but at the same time, several technological barriers require further investigation.

Table 6. Assisted living issues and blockchain solutions.

Ambient Assisted Living Open Issues		Blockchain Solutions	Future Research Directions
Monitoring and Interoperability	IoT sensors integration scalability	Smart contracts for a device to chain integration,	Improve scalability and decrease transactional costs
		transactions based on monitored data	
		Edge off-chain data vs. on-chain transactions	
	Interoperability	single source of truth	Blockchain networks integration, care system legacy application integration
		No data segregation	
Data Management	Data Storage	Decentralized	Decrease storage costs
		Encrypted data	
		Replicated blocks	
		Tamper proof	
	Data sharing	Sharding based on healthcare rules and care paths	Create new care and data-sharing paths for the transition from hospital to home
Security and privacy	Privacy and ethical considerations	safe environment for sharing patient data	Common concept of the data confidentiality, personalization of data disclosure paths
		Informed consent management	
		Creation of a patient's digital identity	

Blockchain scalability is important for integrating the technology into ambient assisted living systems. The monitoring data related to the patient's state and well-being, captured using IoT sensors, must be disseminated through the blockchain network. However, nowadays, the scalability of blockchain networks is low for handling an increasing number of transactions as more people will utilize the platform. The quantity of data to be saved on the blockchain will increase in tandem with the number of transactions on the network. It may cause problems related to the network speed and high costs. It is difficult to assess blockchain performance concerning the integration of IoT monitoring devices in living environments and the storage and dissemination of patient data for remote follow-up. Because the technology is decentralized, there is no benchmark against which to compare performance. However, there are a few methods for assessing performance. One method is to look at how many transactions a blockchain network processes in a particular amount of time. Another technique to assess performance is to look at a blockchain network's average transaction time.

As scalability is a significant issue of the blockchain, managing a high number of transactions is important to gain broad acceptance in ambient assisted living systems. However, there are several solutions for improving scalability. Sharding is one potential solution in which the blockchain is split into several shards, each of which may execute transactions concurrently and can be correlated with the organization and data-sharing procedures in healthcare. It would enable the network to handle a considerably higher volume of transactions without compromising speed or efficiency. Another option is off-chain scaling, which entails shifting part of the data off-chain, and this can be a relevant option even to relieve some of the privacy concerns as the patient monitoring data will be stored at the edge.

In the blockchain, there is a lot of room for privacy. This is relevant for ambient assisted living systems where private data and informed consent must be carefully handled. With blockchain, we may construct a safe and private transactional environment to share patient health records and data. Blockchain, if properly used, has the potential to eliminate fraud and improve transparency. However, ensuring privacy needs further research and development. One of the difficulties is that all parties must have a common concept of the data confidentiality, and new care and data-sharing paths need to be created in the healthcare systems. The creation of a digital identity is another way that blockchain may aid in the enforcement of privacy. This would allow us to choose which personal information is shared and with whom. In an ambient assisted living system, we may, for example, select to share the data partially with our family and not at all with our insurance company. A digital identity would offer us complete control over our personal information, allowing us to guarantee that it is shared only with the people we choose.

Another issue of ambient assisted living systems is the possibility of data leaks and modification. Unauthorized parties may have access to monitoring data if they is not protected adequately. Therefore, data may be encrypted using blockchain. This makes it extremely difficult for anyone who is not allowed to read it and modify it. A decentralized network is what defines a blockchain. It implies that our data are not stored in a single area, making it extremely difficult to be modified. Smart contracts can be created using blockchain technology. Integrated into ambient assisted living systems, they may automatize the IoT devices integration as well as the data processing jobs. As a result, we may designate how and by whom our data can be utilized. If someone tries to use our data in a way we have not approved, the smart contract will prohibit them immediately. The data stay private and safe by utilizing blockchain to encrypt data, build a decentralized network, and construct smart contracts.

Finally, the ability of various systems to communicate and interoperate is important for care and support systems that need to integrate stakeholders across the whole continuum of care. The capacity to exchange currency and data between various blockchain networks is one of the advantages of blockchain interoperability. It may contribute to developing a more integrated and efficient care ecosystem. Another advantage of blockchain interoperability is that it might reduce fragmentation risks. It can assist in guaranteeing that there is a single source of truth and that information is not segregated by allowing multiple blockchain networks to collaborate. Interoperability on the blockchain can let consumers have a more efficient experience. Users may hopefully avoid dealing with numerous distinct care applications by allowing blockchains to interact and integrate. Interoperability across blockchains can also assist in enforcing security. It is possible to uncover potential risks and weaknesses by allowing multiple blockchain networks to share data and information. Interoperability on the blockchain can also assist in cutting expenses. It is possible to prevent duplication of effort and resources.

5. Conclusions

In this paper, we have used the PRISMA methodology to identify, study, and report the relevant state-of-the-art literature around blockchain and its applicability in ambient active living. We have defined inclusion and exclusion criteria, have set several international databases for pooling articles, and finally selected 87 research papers in the qualitative study. As many of the desirable features of ambient assisted living systems may be assured by integrating and using the blockchain technology, we have organized the review to reflect the solutions in relation to the IoT monitoring and integration of environmental sensors, managing and sharing of data, and security and privacy aspects.

The outcome of the study shows that the integration of blockchain with ambient assisted living systems is a hot topic in many of the papers published after 2020. The adoption of remote assistive care was accelerated by the COVID-19 pandemic, and as shown by the qualitative review conducted, blockchain technology can play a significant role in addressing some of the concerns related to ambient assisted living services adoption. Although blockchain technology has the potential to revolutionize the care and ambient assisted living industry, more research is needed to fully understand its implications and applications. Future research includes expanding and replicating existing frameworks, performance, scalability, privacy, and interoperability of blockchain systems in IoT healthcare applications. More studies are needed on the adoption of blockchain in the health and care ecosystem, concentrating on topics such as scalability, costs, creation of new care and data-sharing paths for care transition from hospital to home, governance, and interoperability.

Author Contributions: Conceptualization, A.-I.F. and I.A.; funding acquisition, T.C. and I.A.; investigation, A.-I.F. and I.A.; methodology, A.-I.F. and I.A.; writing—original draft, A.-I.F. and T.C.; writing—review & editing, I.A. and T.C. All authors have read and agreed to the published version of the manuscript.

Funding: This work has been supported by three grants from the Romanian Ministry of Research, Innovation and Digitization, CNCS/CCCDI—UEFISCDI, with co-funding from the AAL Programme (AAL159/2020 H2HCare, AAL264/2021 engAGE, and AAL162/2020 ReMember-Me) and one grant of Romanian Ministry of Research, Innovation and Digitization, CNCS/CCCDI—UEFISCDI (PN-III-P3-3.6-H2020-2020-0031).

Institutional Review Board Statement: Not applicable.

Informed Consent Statement: Not applicable.

Data Availability Statement: Not applicable.

Conflicts of Interest: The authors declare no conflict of interest.

References

1. Cicirelli, G.; Marani, R.; Petitti, A.; Milella, A.; D'Orazio, T. Ambient Assisted Living: A Review of Technologies, Methodologies and Future Perspectives for Healthy Aging of Population. *Sensors* **2021**, *21*, 354. [CrossRef] [PubMed]
2. Anghel, I.; Cioara, T.; Moldovan, D.; Antal, M.; Pop, C.D.; Salomie, I.; Pop, C.B.; Chifu, V.R. Smart Environments and Social Robots for Age-Friendly Integrated Care Services. *Int. J. Environ. Res. Public Health* **2020**, *17*, 3801. [CrossRef] [PubMed]
3. Ramirez, H.; Velastin, S.A.; Meza, I.; Fabregas, E.; Makris, D.; Farias, G. Fall Detection and Activity Recognition Using Human Skeleton Features. *IEEE Access* **2021**, *9*, 33532–33542. [CrossRef]
4. Moldovan, D.; Anghel, I.; Cioara, T.; Salomie, I.; Chifu, V.; Pop, C. Kangaroo Mob Heuristic for Optimizing Features Selection in Learning the Daily Living Activities of People with Alzheimer's. In Proceedings of the 2019 22nd International Conference on Control Systems and Computer Science (CSCS), Bucharest, Romania, 28–30 May 2019; pp. 236–243.

5.   Rochat, J.; Villaverde, A.; Klitzing, H.; Langemyr Larsen, T.; Vogel, M.; Rime, J.; Anghel, I.; Cioara, T.; Lovis, C. *Designing an eHealth Coaching Solution to Improve Transitional Care of Seniors' with Heart Failure: End-User Needs, Studies in Health Technology and Informatics (MIE2021 Articles)*; IOS Press: Amsterdam, The Netherlands, 2021; Volume 281, pp. 530–534. ISBN 978-1-64368-184-9.

6.   Bozdog, I.A.; Daniel-Nicusor, T.; Antal, M.; Antal, C.; Cioara, T.; Anghel, I.; Salomie, I. Human Behavior and Anomaly Detection using Machine Learning and Wearable Sensors. In Proceedings of the 2021 IEEE 17th International Conference on Intelligent Computer Communication and Processing (ICCP), Cluj-Napoca, Romania, 28–30 October 2021; pp. 383–390.

7.   Cioara, T.; Anghel, I.; Salomie, I.; Barakat, L.; Miles, S.; Reidlinger, D.; Taweel, A.; Dobre, C.; Pop, F. Expert system for nutrition care process of older adults. *Future Gener. Comput. Syst.* **2018**, *80*, 368–383. [CrossRef]

8.   Watts, J.; Khojandi, A.; Vasudevan, R.; Nahab, F.B.; Ramdhani, R.A. Improving Medication Regimen Recommendation for Parkinson's Disease Using Sensor Technology. *Sensors* **2021**, *21*, 3553. [CrossRef]

9.   Sărătean, T.; Antal, M.; Pop, C.; Cioara, T.; Anghel, I.; Salomie, I. A Physiotheraphy Coaching System based on Kinect Sensor. In Proceedings of the 2020 IEEE 16th International Conference on Intelligent Computer Communication and Processing (ICCP), Cluj-Napoca, Romania, 3–5 September 2020; pp. 535–540.

10.   Ismail, L.; Materwala, H.; Karduck, A.P.; Adem, A. Requirements of Health Data Management Systems for Biomedical Care and Research: Scoping Review. *J. Med. Internet Res.* **2020**, *22*, e17508. [CrossRef]

11.   Cubo, J.; Nieto, A.; Pimentel, E. A Cloud-Based Internet of Things Platform for Ambient Assisted Living. *Sensors* **2014**, *14*, 14070–14105. [CrossRef]

12.   Xu, M.; Chen, X.; Kou, G. A systematic review of blockchain. *Financ. Innov.* **2019**, *5*, 27. [CrossRef]

13.   Husnoo, M.; Anwar, A.; Chakrabortty, R.; Doss, R.; Ryan, M. Differential Privacy for IoT-Enabled Critical Infrastructure: A Comprehensive Survey. *IEEE Access* **2021**, *9*, 153276–153304. [CrossRef]

14.   Kumar, R.; Tripathi, R. Towards design and implementation of security and privacy framework for Internet of Medical Things (IoMT) by leveraging blockchain and IPFS technology. *J. Supercomput.* **2021**, *77*, 7916–7955. [CrossRef]

15.   Esfahani, M.; Ghahfarokhi, B.; Borujeni, S. End-to-end privacy preserving scheme for IoT-based healthcare systems. *Wirel. Netw.* **2021**, *27*, 4009–4037. [CrossRef]

16.   Aceto, G.; Persico, V.; Pescapé, A. A Survey on Information and Communication Technologies for Industry 4.0: State-of-the-Art, Taxonomies, Perspectives, and Challenges. *IEEE Commun. Surv. Tutor.* **2019**, *4*, 3467–3501. [CrossRef]

17.   Kashani, M.; Madanipour, M.; Nikravan, M.; Asghari, P.; Mahdipour, E. A systematic review of IoT in healthcare: Applications, techniques, and trends. *J. Netw. Comput. Appl.* **2021**, *192*, 103164. [CrossRef]

18.   Krishnamoorthy, S.; Dua, A.; Gupta, S. Role of emerging technologies in future IoT-driven Healthcare 4.0 technologies: A survey, current challenges and future directions. *J. Ambient. Intell. Humaniz. Comput.* **2021**, *12*, 1–47. [CrossRef]

19.   Poongodi, M.; Sharma, A.; Hamdi, M.; Maode, M.; Chilamkurti, N. Smart healthcare in smart cities: Wireless patient monitoring system using IoT. *J. Supercomput.* **2021**, *77*, 12230–12255. [CrossRef]

20.   Calvillo-Arbizu, J.; Roman-Martinez, I.; Reina-Tosina, J. Internet of things in health: Requirements, issues, and gaps. *Comput. Methods Programs Biomed.* **2021**, *208*, 106231. [CrossRef]

21.   Mamdouh, M.; Awad, A.; Khalaf, A.; Hamed, H. Authentication and Identity Management of IoHT Devices: Achievements, Challenges, and Future Directions. *Comput. Secur.* **2021**, *111*, 102491. [CrossRef]

22.   Satamraju, K.; Malarkodi, B. A decentralized framework for device authentication and data security in the next generation internet of medical things. *Comput. Commun.* **2021**, *180*, 146–160. [CrossRef]

23.   Antal, C.; Cioara, T.; Anghel, I.; Antal, M.; Salomie, I. Distributed Ledger Technology Review and Decentralized Applications Development Guidelines. *Future Internet* **2021**, *13*, 62. [CrossRef]

24.   Ray, P.; Dash, D.; Salah, K.; Kumar, N. Blockchain for IoT-Based Healthcare: Background, Consensus, Platforms, and Use Cases. *IEEE Syst. J.* **2021**, *15*, 85–94. [CrossRef]

25.   Hussien, H.; Yasin, S.; Udzir, N.; Ninggal, M.; Salman, S. Blockchain technology in the healthcare industry: Trends and opportunities. *J. Ind. Inf. Integr.* **2021**, *22*, 100217. [CrossRef]

26.   Bouras, M.A.; Lu, Q.; Zhang, F.; Wan, Y.; Zhang, T.; Ning, H. Distributed Ledger Technology for eHealth Identity Privacy: State of The Art and Future Perspective. *Sensors* **2020**, *20*, 483. [CrossRef] [PubMed]

27.   ElRahman, S.; Alluhaidan, A. Blockchain technology and IoT-edge framework for sharing healthcare services. *Soft Comput.* **2021**, *25*, 13753–13777. [CrossRef]

28.   Bataineh, M.; Mardini, W.; Khamayseh, Y.; Yassein, M. Novel and Secure Blockchain Framework for Health Applications in IoT. *IEEE Access* **2022**, *10*, 14914–14926. [CrossRef]

29.   Dwivedi, A.; Singh, R.; Ghosh, U.; Mukkamala, R.; Tolba, A.; Said, O. Privacy preserving authentication system based on non-interactive zero knowledge proof suitable for Internet of Things. *J. Ambient. Intell. Humaniz. Comput.* **2021**, *12*, 1–11. [CrossRef]

30.   Hau, Y.; Chang, M. A Quantitative and Qualitative Review on the Main Research Streams Regarding Blockchain Technology in Healthcare. *Healthcare* **2021**, *9*, 247. [CrossRef]

31.   Nasir, M.; Arshad, J.; Khan, M.; Fatima, M.; Salah, K.; Jayaraman, R. Scalable blockchains—A systematic review. *Future Gener. Comput. Syst. Int. J. eScience* **2022**, *126*, 136–162. [CrossRef]

32.   Haque, A.; Islam, A.; Hyrynsalmi, S.; Naqvi, B.; Smolander, K. GDPR Compliant Blockchains—A Systematic Literature Review. *IEEE Access* **2021**, *9*, 50593–50606. [CrossRef]

33. Oláh, J.; Krisán, E.; Kiss, A.; Lakner, Z.; Popp, J. PRISMA Statement for Reporting Literature Searches in Systematic Reviews of the Bioethanol Sector. *Energies* **2020**, *13*, 2323. [CrossRef]

34. Web of Science Platform. Available online: https://www.webofscience.com/wos/woscc/basic-search (accessed on 1 March 2022).

35. Kakarlapudi, P.; Mahmoud, Q. A Systematic Review of Blockchain for Consent Management. *Healthcare* **2021**, *9*, 137. [CrossRef]

36. Farahani, B.; Firouzi, F.; Luecking, M. The convergence of IoT and distributed ledger technologies (DLT): Opportunities, challenges, and solutions. *J. Netw. Comput. Appl.* **2021**, *177*, 102936. [CrossRef]

37. Sanka, A.; Irfan, M.; Huang, I.; Cheung, R. A survey of breakthrough in blockchain technology: Adoptions, applications, challenges and future research. *Comput. Commun.* **2021**, *169*, 179–201. [CrossRef]

38. Egala, B.; Pradhan, A.; Badarla, V.; Mohanty, S. Fortified-Chain: A Blockchain-Based Framework for Security and Privacy-Assured Internet of Medical Things with Effective Access Control. *IEEE Internet Things J.* **2021**, *8*, 11717–11731. [CrossRef]

39. Nguyen, D.; Pathirana, P.; Ding, M.; Seneviratne, A. BEdgeHealth: A Decentralized Architecture for Edge-Based IoMT Networks Using Blockchain. *IEEE Internet Things J.* **2021**, *8*, 11743–11757. [CrossRef]

40. Gul, M.; Subramanian, B.; Paul, A.; Kim, J. Blockchain for public health care in smart society. *Microprocess. Microsyst.* **2021**, *80*, 103524. [CrossRef]

41. Bhuiyan, M.; Rahman, M.; Billah, M.; Saha, D. Internet of Things (IoT): A Review of Its Enabling Technologies in Healthcare Applications, Standards Protocols, Security, and Market Opportunities. *IEEE Internet Things J.* **2021**, *8*, 10474–10498. [CrossRef]

42. Al Sadawi, A.; Hassan, M.; Ndiaye, M. A Survey on the Integration of Blockchain with IoT to Enhance Performance and Eliminate Challenges. *IEEE Access* **2021**, *9*, 54497. [CrossRef]

43. Arul, R.; Alroobaea, R.; Tariq, U.; Almulihi, A.; Alharithi, F.; Shoaib, U. IoT-enabled healthcare systems using block chain-dependent adaptable services. *Pers. Ubiquitous Comput.* **2021**, *25*, 1–15. [CrossRef]

44. Abbas, A.; Alroobaea, R.; Krichen, M.; Rubaiee, S.; Vimal, S.; Almansour, F. Blockchain-assisted secured data management framework for health information analysis based on Internet of Medical Things. *Pers. Ubiquitous Comput.* **2021**, *25*, 1–14. [CrossRef]

45. Attarian, R.; Hashemi, S. An anonymity communication protocol for security and privacy of clients in IoT-based mobile health transactions. *Comput. Netw.* **2021**, *190*, 107976. [CrossRef]

46. Hossein, K.; Esmaeili, M.; Dargahi, T.; Khonsari, A.; Conti, M. BCHealth: A Novel Blockchain-based Privacy-Preserving Architecture for IoT Healthcare Applications. *Comput. Commun.* **2021**, *180*, 31–47. [CrossRef]

47. Abdelmaboud, A.; Ahmed, A.; Abaker, M.; Eisa, T.; Albasheer, H.; Ghorashi, S.; Karim, F. Blockchain for IoT Applications: Taxonomy, Platforms, Recent Advances, Challenges and Future Research Directions. *Electronics* **2022**, *11*, 630. [CrossRef]

48. Sharma, P.; Jindal, R.; Borah, M. A review of smart contract-based platforms, applications, and challenges. *Clust. Comput.* **2021**. [CrossRef]

49. Zulkifl, Z.; Khan, F.; Tahir, S.; Afzal, M.; Iqbal, W.; Rehman, A.; Saeed, S.; Almuhaideb, A. FBASHI: Fuzzy and Blockchain-Based Adaptive Security for Healthcare IoTs. *IEEE Access* **2022**, *10*, 15644–15656. [CrossRef]

50. Kakarlapudi, P.; Mahmoud, Q. Design and Development of a Blockchain-Based System for Private Data Management. *Electronics* **2021**, *10*, 3131. [CrossRef]

51. Swathi, P.; Venkatesan, M. Scalability improvement and analysis of permissioned-blockchain. *ICT Express* **2021**, *7*, 283–289.

52. Shukla, S.; Thakur, S.; Hussain, S.; Breslin, J.; Jameel, S. Identification and Authentication in Healthcare Internet-of-Things Using Integrated Fog Computing Based Blockchain Model. *Internet Things* **2021**, *15*, 100422. [CrossRef]

53. Chinaei, M.; Gharakheili, H.; Sivaraman, V. Optimal Witnessing of Healthcare IoT Data Using Blockchain Logging Contract. *IEEE Internet Things J.* **2021**, *8*, 10117–10130. [CrossRef]

54. Jolfaei, A.; Aghili, S.; Singelee, D. A Survey on Blockchain-Based IoMT Systems: Towards Scalability. *IEEE Access* **2021**, *9*, 148948–148975. [CrossRef]

55. Girardi, F.; De Gennaro, G.; Colizzi, L.; Convertini, N. Improving the Healthcare Effectiveness: The Possible Role of EHR, IoMT and Blockchain. *Electronics* **2020**, *9*, 884. [CrossRef]

56. Aujla, G.S.; Singh, M.; Bose, A.; Kumar, N.; Han, G.; Buyya, R. BlockSDN: Blockchain-as-a-Service for Software Defined Networking in Smart City Applications. *IEEE Netw.* **2020**, *34*, 89–91. [CrossRef]

57. Calvaresi, D.; Dubovitskaya, A.; Calbimonte, J.P.; Taveter, K.; Schumacher, M. Multi-Agent Systems and Blockchain: Results from a Systematic Literature Review. In *Advances in Practical Applications of Agents, Multi-Agent Systems, and Complexity: The PAAMS Collection*; Springer: Cham, Switzerland, 2018.

58. Mendes, D.; Rodrigues, I.; Fonseca, C.; Lopes, M.; García-Alonso, J.M.; Berrocal, J. Anonymized Distributed PHR Using Blockchain for Openness and Non-Repudiation Guarantee. In *Decision Support Systems and Education: Help and Support in Healthcare*; IOP Press: Bristol, UK, 2018.

59. Spinsante, S.; Poli, A.; Mongay Batalla, J.; Krawiec, P.; Dobre, C.; Băjenaru, L.; Mavromoustakis, C.X.; Costantinou, C.S.; Molan, G.; Herghelegiu, A.M.; et al. Clinically-validated technologies for assisted living The vINCI project. *J. Ambient. Intell. Humaniz. Comput.* **2021**, *12*, 1–22. [CrossRef]

60. Velmovitsky, P.E.; Souza, P.A.D.S.E.; Vaillancourt, H.; Donovska, T.; Teague, J.; Morita, P.P. A Blockchain-Based Consent Platform for Active Assisted Living: Modeling Study and Conceptual Framework. *J. Med. Internet Res.* **2020**, *22*, e20832. [CrossRef] [PubMed]

61. Tripathi, G.; Abdul Ahad, M.; Paiva, S. SMS: A Secure Healthcare Model for Smart Cities. *Electronics* **2020**, *9*, 1135. [CrossRef]
62. Ali, M.S.; Vecchio, M.; Putra, G.D.; Kanhere, S.S.; Antonelli, F. A Decentralized Peer-to-Peer Remote Health Monitoring System. *Sensors* **2020**, *20*, 1656. [CrossRef]
63. Ahmad, R.W.; Salah, K.; Jayaraman, R.; Yaqoob, I.; Ellahham, S.; Omar, M. The role of blockchain technology in telehealth and telemedicine. *Int. J. Med. Inform.* **2021**, *148*, 104399. [CrossRef]
64. Uddin, M.A.; Stranieri, A.; Gondal, I.; Balasubramanian, V. Continuous Patient Monitoring with a Patient Centric Agent: A Block Architecture. *IEEE Access* **2018**, *6*, 32700–32726. [CrossRef]
65. Hathaliya, J.; Sharma, P.; Tanwar, S.; Gupta, R. Blockchain-based Remote Patient Monitoring in Healthcare 4.0. In Proceedings of the 2019 IEEE 9th International Conference on Advanced Computing (IACC 2019), Tiruchirapalli, India, 13–14 December 2019.
66. Elangovan, D.; Long, C.S.; Bakrin, F.S.; Tan, C.S.; Goh, K.W.; Yeoh, S.F.; Loy, M.J.; Hussain, Z.; Lee, K.S.; Idris, A.C.; et al. The Use of Blockchain Technology in the Health Care Sector: Systematic Review. *JMIR Med. Inform.* **2022**, *10*, e17278. [CrossRef]
67. Javed, I.T.; Alharbi, F.; Bellaj, B.; Margaria, T.; Crespi, N.; Qureshi, K.N. Health-ID: A Blockchain-Based Decentralized Identity Management for Remote Healthcare. *Healthcare* **2021**, *9*, 712. [CrossRef]
68. Durneva, P.; Cousins, K.; Chen, M. The Current State of Research, Challenges, and Future Research Directions of Blockchain Technology in Patient Care: Systematic Review. *J. Med. Internet Res.* **2020**, *22*, e18619. [CrossRef]
69. Indumathi, J.; Shankar, A.; Ghalib, M.R.; Gitanjali, J.; Hua, Q.; Wen, Z.; Qi, X. Block Chain Based Internet of Medical Things for Uninterrupted, Ubiquitous, User-Friendly, Unflappable, Unblemished, Unlimited Health Care Services (BC IoMT U^6 HCS). *IEEE Access* **2020**, *8*, 216856–216872. [CrossRef]
70. Fernández-Caramés, T.M.; Froiz-Míguez, I.; Blanco-Novoa, O.; Fraga-Lamas, P. Enabling the Internet of Mobile Crowdsourcing Health Things: A Mobile Fog Computing, Blockchain and IoT Based Continuous Glucose Monitoring System for Diabetes Mellitus Research and Care. *Sensors* **2019**, *19*, 3319. [CrossRef] [PubMed]
71. Tandon, A.; Dhir, A.; Islam, A.N.; Mäntymäki, M. Blockchain in healthcare: A systematic literature review, synthesizing framework and future research agenda. *Comput. Ind.* **2020**, *122*, 103290. [CrossRef]
72. Abdellatif, A.A.; Al-Marridi, A.Z.; Mohamed, A.; Erbad, A.; Chiasserini, C.F.; Refaey, A. ssHealth: Toward Secure, Blockchain-Enabled Healthcare Systems. *IEEE Network* **2020**, *34*, 312–319. [CrossRef]
73. Zheng, X.; Sun, S.; Mukkamala, R.R.; Vatrapu, R.; Ordieres-Meré, J. Accelerating Health Data Sharing: A Solution Based on the Internet of Things and Distributed Ledger Technologies. *J. Med. Internet Res.* **2019**, *21*, e13583. [CrossRef]
74. Wang, J.; Han, K.; Alexandridis, A.; Chen, Z.; Zilic, Z.; Pang, Y.; Jeon, G.; Piccialli, F. A blockchain-based eHealthcare system interoperating with WBANs. *Future Gener. Comput. Syst.* **2020**, *110*, 675–685. [CrossRef]
75. Ejaz, M.; Kumar, T.; Kovacevic, I.; Ylianttila, M.; Harjula, E. Health-BlockEdge: Blockchain-Edge Framework for Reliable Low-Latency Digital Healthcare Applications. *Sensors* **2021**, *21*, 2502. [CrossRef]
76. Zhang, P.; White, J.; Schmidt, D.C.; Lenz, G.; Rosenbloom, S.T. FHIRChain: Applying Blockchain to Securely and Scalably Share Clinical Data. *Comput. Struct. Biotechnol. J.* **2018**, *16*, 267–278. [CrossRef]
77. Mahmud, H.; Rahman, T. An Application of blockchain to securely acquire, diagnose and share clinical data through smartphone. *Peer-to-Peer Netw. Appl.* **2021**, *14*, 3758–3777. [CrossRef]
78. Ichikawa, D.; Kashiyama, M.; Ueno, T. Tamper-Resistant Mobile Health Using Blockchain Technology. *JMIR mHealth uHealth* **2017**, *5*, e7938. [CrossRef]
79. Taralunga, D.D.; Florea, B.C. A Blockchain-Enabled Framework for mHealth Systems. *Sensors* **2021**, *21*, 2828. [CrossRef] [PubMed]
80. Li, C.T.; Shih, D.H.; Wang, C.C.; Chen, C.L.; Lee, C.C. A Blockchain Based Data Aggregation and Group Authentication Scheme for Electronic Medical System. *IEEE Access* **2020**, *8*, 173904–173917. [CrossRef]
81. Jamil, F.; Ahmad, S.; Iqbal, N.; Kim, D.H. Towards a Remote Monitoring of Patient Vital Signs Based on IoT-Based Blockchain Integrity Management Platforms in Smart Hospitals. *Sensors* **2020**, *20*, 2195. [CrossRef] [PubMed]
82. Malamas, V.; Kotzanikolaou, P.; Dasaklis, T.K.; Burmester, M. A Hierarchical Multi Blockchain for Fine Grained Access to Medical Data. *IEEE Access* **2020**, *8*, 134393–134412. [CrossRef]
83. Fatoum, H.; Hanna, S.; Halamka, J.D.; Sicker, D.C.; Spangenberg, P.; Hashmi, S.K. Blockchain Integration with Digital Technology and the Future of Health Care Ecosystems: Systematic Review. *J. Med. Internet Res.* **2021**, *23*, e19846. [CrossRef]
84. Ali, O.; Jaradat, A.; Kulakli, A.; Abuhalimeh, A. A Comparative Study: Blockchain Technology Utilization Benefits, Challenges and Functionalities. *IEEE Access* **2021**, *9*, 12730–12749. [CrossRef]
85. Radanović, I.; Likić, R. Opportunities for Use of Blockchain Technology in Medicine. *Appl. Health Econ. Health Policy* **2018**, *16*, 583–590. [CrossRef]
86. Zhang, P.; Schmidt, D.C.; White, J.; Lenz, G. Blockchain Technology Use Cases in Healthcare. In *Blockchain Technology: Platforms, Tools and Use Cases*; AP Publishing: Cascina, Italy, 2018.
87. Vazirani, A.A.; O'Donoghue, O.; Brindley, D.; Meinert, E. Blockchain vehicles for efficient Medical Record management. *NPJ Digit. Med.* **2020**, *3*, 1–5. [CrossRef]
88. Zhuang, Y.; Sheets, L.R.; Chen, Y.W.; Shae, Z.Y.; Tsai, J.J.; Shyu, C.R. A Patient-Centric Health Information Exchange Framework Using Blockchain Technology. *IEEE J. Biomed. Health Inform.* **2020**, *24*, 2169–2176. [CrossRef]
89. Fang, H.S.A.; Tan, T.H.; Tan, Y.F.C.; Tan, C.J.M. Blockchain Personal Health Records: Systematic Review. *J. Med. Internet Res.* **2021**, *23*, e25094. [CrossRef]

90. Hickman, C.F.L.; Alshubbar, H.; Chambost, J.; Jacques, C.; Pena, C.A.; Drakeley, A.; Freour, T. Data sharing: Using blockchain and decentralized data technologies to unlock the potential of artificial intelligence: What can assisted reproduction learn from other areas of medicine? *Fertil. Steril.* **2020**, *114*, 927–933. [CrossRef] [PubMed]

91. Rahman, A.; Rashid, M.; Barnes, S.; Hossain, M.S.; Hassanain, E.; Guizani, M. An IoT and Blockchain-Based Multi-Sensory In-Home Quality of Life Framework for Cancer Patients. In Proceedings of the 2019 15th International Wireless Communications & Mobile Computing Conference (IWCMC), Tangier, Morocco, 24–28 June 2019.

92. Shae, Z.; Tsai, J.J. On the Design of a Blockchain Platform for Clinical Trial and Precision Medicine. In Proceedings of the 2017 IEEE 37th International Conference on Distributed Computing Systems (ICDCS 2017), Atlanta, GA, USA, 5 June 2017.

93. Lee, H.A.; Kung, H.H.; Udayasankaran, J.G.; Kijsanayotin, B.; Marcelo, A.B.; Chao, L.R.; Hsu, C.Y. An Architecture and Management Platform for Blockchain-Based Personal Health Record Exchange: Development and Usability Study. *J. Med. Internet Res.* **2020**, *6*, e16748. [CrossRef] [PubMed]

94. Rahman, A.; Rashid, M.; Le Kernec, J.; Philippe, B.; Barnes, S.J.; Fioranelli, F.; Yang, S.; Romain, O.; Abbasi, Q.H.; Loukas, G.; et al. A Secure Occupational Therapy Framework for Monitoring Cancer Patients' Quality of Life. *Sensors* **2019**, *19*, 5258. [CrossRef] [PubMed]

95. Dammak, B.; Turki, M.; Cheikhrouhou, S.; Baklouti, M.; Mars, R.; Dhahbi, A. LoRaChainCare: An IoT Architecture Integrating Blockchain and LoRa Network for Personal Health Care Data Monitoring. *Sensors* **2022**, *22*, 1497. [CrossRef] [PubMed]

96. El Majdoubi, D.; El Bakkali, H.; Sadki, S. SmartMedChain: A Blockchain-Based Privacy-Preserving Smart Healthcare Framework. *J. Healthc. Eng.* **2021**, *2021*, 4145512. [CrossRef] [PubMed]

97. Akkaoui, R.; Hei, X.; Cheng, W. EdgeMediChain: A Hybrid Edge Blockchain-Based Framework for Health Data Exchange. *IEEE Access* **2020**, *8*, 113467–113486. [CrossRef]

98. Duhayyim, M.A.; Al-Wesabi, F.N.; Marzouk, R.; Musa, A.I.A.; Negm, N.; Hilal, A.M.; Hamza, M.A.; Rizwanullah, M. Integration of Fog Computing for Health Record Management Using Blockchain Technology. *Comput. Mater. Contin.* **2022**, *71*, 4135–4149. [CrossRef]

99. Sengupta, A.; Subramanian, H. User Control of Personal mHealth Data Using a Mobile Blockchain App: Design Science Perspective. *JMIR mHealth uHealth* **2022**, *10*, e32104. [CrossRef]

100. Sylla, T.; Mendiboure, L.; Chalouf, M.A.; Krief, F. Blockchain-Based Context-Aware Authorization Management as a Service in IoT. *Sensors* **2021**, *21*, 7656. [CrossRef]

101. Chen, W.; Zhu, S.; Li, J.; Wu, J.; Chen, C.-L.; Deng, Y.-Y. Authorized Shared Electronic Medical Record System with Proxy Re-Encryption and Blockchain Technology. *Sensors* **2021**, *21*, 7765. [CrossRef]

MDPI

St. Alban-Anlage 66

4052 Basel

Switzerland

Tel. +41 61 683 77 34

Fax +41 61 302 89 18

www.mdpi.com

Future Internet Editorial Office

E-mail: futureinternet@mdpi.com

www.mdpi.com/journal/futureinternet